Progress in IS

"PROGRESS in IS" encompasses the various areas of Information Systems in theory and practice, presenting cutting-edge advances in the field. It is aimed especially at researchers, doctoral students, and advanced practitioners. The series features both research monographs that make substantial contributions to our state of knowledge and handbooks and other edited volumes, in which a team of experts is organized by one or more leading authorities to write individual chapters on various aspects of the topic. "PROGRESS in IS" is edited by a global team of leading IS experts. The editorial board expressly welcomes new members to this group. Individual volumes in this series are supported by a minimum of two members of the editorial board, and a code of conduct mandatory for all members of the board ensures the quality and cutting-edge nature of the titles published under this series.

More information about this series at https://link.springer.com/bookseries/10440

Gustaf Juell-Skielse • Ida Lindgren •
Maria Åkesson
Editors

Service Automation in the Public Sector

Concepts, Empirical Examples and Challenges

Springer

Editors
Gustaf Juell-Skielse
University of Borås
Borås, Sweden

Ida Lindgren
Linköping University
Linköping, Sweden

Maria Åkesson
Halmstad University
Halmstad, Sweden

ISSN 2196-8705 ISSN 2196-8713 (electronic)
Progress in IS
ISBN 978-3-030-92643-4 ISBN 978-3-030-92644-1 (eBook)
https://doi.org/10.1007/978-3-030-92644-1

This Springer imprint is published by the registered company Springer Nature Switzerland AG
The registered company address is: Gewerbestrasse 11, 6330 Cham, Switzerland

Preface

With great satisfaction, we now present this book on service automation in public organizations. This book was written because we experience a strong interest from our partners in the public sector in issues about artificial intelligence (AI), robotic process automation (RPA), and how automation contributes to the digital transformation of their organizations. The first thoughts of a book came when we realized that we worked with the same types of questions and case studies about RPA in different municipalities. A book could be something to gather around and a way to form a community of researchers interested in the use of RPA in the public sector. Therefore, in connection with The Scandinavian Workshop on E-Government (SWEG 2020), researchers in digital government met to discuss the possibilities of collaborating on robotic processes automation issues in public organizations. However, we soon saw a need to expand the scope to deal with service automation in general rather than specifically RPA.

Our partners in the public sector had many questions about the development of new organizational skills, benefits, implementation, and challenges. As usual when an interest turns into a trend, there are great risks of excessive optimism. There is also a cause for concern when technologies that have first emerged in the private sector are uncritically imported into the public sector without considerations of possible differences between the sectors. Against this backdrop, we saw the need to describe, problematize, and analyze what an increase of automation in the public sector can entail. We furthermore wanted to address automation from different perspectives and acknowledge the particularities of the public sector.

The outbreak of COVID-19 unfortunately delayed the start somewhat, but a call for chapters was presented in the autumn of 2020 and a website[1] was set up for communication with interested researchers. The call attracted attention from an international crowd of researchers from multiple fields and contexts. During the winter and spring of 2021, a series of chapter development workshops were held to

[1] https://publicserviceautomation.com

support authors in developing their chapter ideas. During the following summer and autumn, double-blind peer reviews of the submitted chapters were conducted. In total, twenty submissions were examined, which resulted in ten chapters. Additionally, two chapters were written at the invitation of the editors.

We have divided the selected chapters into three main parts: conceptualization, applications, and implementation challenges. Conceptualization aims to clarify the core concepts related to public service automation. Applications presents empirical examples of automation in public organizations. Implementation Challenges includes chapters identifying and discussing challenges that can arise from the implementation of automation technologies in the public sector service. The Editorial in first chapter briefly introduces the chapters in each of these three parts, and based on the lessons learned presents calls for further research on public service automation.

We thank Christian Rauscher at Springer Nature for encouraging and supporting the initial idea for this book, and Ramya Prakash and Jialin Yan for coordinating the work and supporting us during the production of the book. We also wish to thank the researchers who submitted chapters. Finally, we are grateful to our colleagues and friends who supported this book project by serving on its editorial board and who dedicated much of their time in reviewing and providing feedback on the submitted chapters.

We hope that you will enjoy reading the book and invite you to contact us for questions, feedback, and discussions.

Editors

Borås, Sweden — Gustaf Juell-Skielse
Linköping, Sweden — Ida Lindgren
Halmstad, Sweden — Maria Åkesson

Editorial Board

Contents

Part I
Editorial

Towards Service Automation in Public Organizations

Gustaf Juell-Skielse, Ida Lindgren, and Maria Åkesson

1 Introduction

Public sector organizations worldwide are under pressure to operate more efficiently, serve their citizens well, and provide a good working environment for their employees. With limited budgets, these goals are challenging and require new and innovative ways of organizing public sector operations. Inspired by successful implementations of automation technologies in the private sector, such as robotic process automation (RPA) and services based on artificial intelligence (AI), public sector organizations are currently exploring whether similar solutions can bring improvements also for public sector operations.

Automation of services and administrative processes has re-emerged as a popular theme also in the research discourse on digital government and public administration (Lindgren et al., 2019). Automation of structured and high-volume routine tasks is discussed as a means to shorten lead times and reduce costs for manual labor in public service delivery (Wirtz & Müller, 2019). If and to what extent automation of services and interconnected work practices and routines will lead to the expected gains for public organizations is still an open question. This question has served as a starting point and inspiration for this book.

G. Juell-Skielse (✉)
University of Borås, Borås, Sweden
e-mail: gustaf.juell-skielse@hb.se

I. Lindgren
Linköping University, Linköping, Sweden
e-mail: ida.lindgren@liu.se

M. Åkesson
Halmstad University, Halmstad, Sweden
e-mail: maria.akesson@hh.se

G. Juell-Skielse et al. (eds.), *Service Automation in the Public Sector*, Progress in IS,
https://doi.org/10.1007/978-3-030-92644-1_1

A current example of a digital technology used for automation of services and work processes in public organizations is RPA, referring to software that can be programmed to interact with user interfaces of information systems in a way that imitate human end-users. This type of software is designed to perform repetitive tasks quickly and is an example of the so-called lightweight information technology (Bygstad, 2012) that can be used to integrate information from various kinds of information systems. RPA was originally oriented towards business with fast and user-driven implementations and has in the private sector been found to increase customer satisfaction, financial performance, and process compliance (Lacity & Willcocks, 2021). Also, RPA has been found to free employees from repetitive and monotonous work so that they become available for more critical and value adding tasks in the organization (Suri et al., 2017; Bhatnagar, 2020). Based on these experiences from the business context, researchers and public sector officials alike see potential in using RPA also in public organizations. Hopes are that service automation using RPA will lead to lower costs, more efficient processes, fewer errors, time savings, and improved service quality of citizen services (Ranerup & Henriksen, 2019).

In the current discourse on the utility of RPA for public sector organizations, RPA is discussed as a silver bullet that can resolve the challenges faced by public sector organizations, e.g. challenges related to inefficiency in public administration, decreasing funds for public service delivery, working environment issues for public officials, and difficulties in attracting and recruiting personnel with the right competence. However, how well automation technologies such as RPA can perform in relation to public service delivery remains an empirical question; we still lack empirical evidence for the success and suitability of automation technologies, such as RPA, for public service delivery. There is yet limited research on service automation in the public sector, and there are only a few studies available that target automation technologies involving RPA or AI elements in public sector organizations (e.g., Wihlborg et al., 2016; Ranerup & Henriksen, 2019, 2020; Toll et al., 2021; Lindgren et al., 2021).

This title aims at presenting the latest advancements and findings from research on service automation in public sector organizations. Examples of service automation include, but are not limited to, Robotic Process Automation, Cognitive RPA, and various AI technologies. Spread across eleven chapters, this book brings forward conceptual- and empirical work from social- as well as technical perspectives. The targeted audience includes researchers interested in digital government and digital transformation; practitioners in public sector organizations working actively with automation; policy makers at local, regional, national, or international government levels; university students and professors from research disciplines concerned with people's and organizations' use of digital technologies; as well as ICT industry experts engaged in public sector information system, software design, and deployment projects related to automation. As such, this book is built on an ambition to aid and inspire researchers and practitioners to advance their knowledge on service automation in public organizations, as well as to provide a foundation for policy development and future research.

2 Understanding Public Sector Service Automation

The research presented in this book investigates public sector service automation from several perspectives. We have organized the content in three thematic sections that deal with the *conceptualization* of this phenomenon (Sect. 2.1), empirical examples of automation *applications* in public organizations (Sect. 2.2), and *implementation challenges* that can arise from public sector service automation (Sect. 2.3). Together, these three sections further our understanding of public sector service automation as a concept, as manifested in practice, and its associated opportunities and challenges.

2.1 Conceptualization of Public Sector Service Automation

Automation in public sector organizations is not a new phenomenon; however, lately software robots have emerged as a new way to automate larger portions of work processes. In light of this new wave of organizational development, Goldkuhl identifies a need for clarifying the concept of *process automation.* He performs a meta-study on several ongoing research projects and investigates basic concepts from a linguistic and ontological perspective. The results include a linguistic analysis of automation and a structure of key practices related to process automation. It provides a shared conceptual and linguistic ground for researchers in process automation in public sector organizations.

In the subsequent chapter, Roehl delves deeper into the case handling practices, illustrated by Goldkuhl, and takes a closer look at different decision situations where automation technology is used. Roehl presents a functional classification of six ideal types of decision situations with various degree of automation using algorithmic systems. The *typology* clarifies, for each of the six ideal decision situations, how decision authority is distributed between civil servants and algorithmic systems on the other. Through the typology, Roehl focuses on civil servants use of technology and sheds light on differences in organizational practices depending on different levels of human involvement and decision authority entrusted with technology. The typology informs practitioners and policy makers about actual work practices and how these are reflected in the design of IT systems.

In a similar vein, Juell-Skielse, Balasuriya, Güner, and Han identify a lack of conceptualization of cognitive robotic process automation and how it affects public organizations' dynamic IT capabilities. Juell-Skielse et al. offer a definition of cognitive RPA and depict it as an open system. Also, the authors offer a set of propositions for how an extended notion of RPA affects dynamic IT capabilities in public sector organizations. They provide testable hypotheses to researchers and inform practitioners of how RPA affects their organizational capabilities.

Together these three chapters provide solid foundations for defining practices related to process automation, distribution of decision authority between civil servants and algorithmic systems as well as extending the notion of robotic process automation to include cognitive artificial intelligence.

2.2 Applications of Public Sector Service Automation

Automation of work has re-emerged as a popular theme in public administration, where suppliers of automation technologies and AI-based services promise fast and bias-free handling of citizen data. In parallel, policy makers promote increased digitalization and automation as a way of increasing efficiency in administrative routines. Two policy areas in which RPA and other automation techniques have gained attention are *human resource management* (HRM) and *social work*, both of which are associated with large volumes of paperwork and rule-based decision-making. However, they also include social interactions with employees/clients, professional work, and discretionary practices. Three chapters of this book provide insights into the use of RPA in HRM and social work.

First, based on a literature review, Persson and Wallo provide an overview of RPA use within the field of HRM and relate this to public service values. Their review indicates that automating IT systems can be an effective way to creating efficiency in administrative and information-based services provided by HRM. But when it comes to more innovation-oriented services, RPA seems to have few if any benefits. Many organizations seem to utilize e-HRM more for an automating approach that focuses primarily on increasing administrative efficiency rather than supporting strategic human capital management processes.

As a complement, Gustafsson illustrates how algorithms-supported decision-making in social work requires new and cross-jurisdictional governance and coordination mechanisms, as well as new institutional arrangements. As such, RPA implementation in social work creates tensions, ambiguities, and conflicts that call for public organizations to re-examine current practices and core services in relation to democratic goals.

On a similar note, Ranerup and Svensson provide a qualitative study of how RPA is discussed and implemented in social work in Sweden. Their results call for future studies on the discretion of caseworkers in view of the greater experiences of RPA, as well as the influence of new laws that might, or might not, increase the potential of utilizing RPA in local government contexts like social work.

Last, Neu, Benke, Müller, and Fettke illuminate automation from a slightly different perspective by highlighting the use of RPA for cross-border business processes. Their study shows that public organizations can engender citizen trust in AI service agents by enhancing their sense of security and data protection, as well as the perceived quality of advice.

All four chapters give important insights on how automation of work and services can be manifested in practice and show how different values can be created using automation technologies, as well as highlighting challenges that can arise from the use of automation.

2.3 *Implementation Challenges of Public Sector Service Automation*

The last section of chapters provides additional examples of automation applications in public organizations but focus on what challenges public organizations can face during the phase in which they implement automation technologies.

The study by Güner, Han, and Juell-Skielse shows that only a few early adopters have implemented RPA in the public sector, and that RPA development and use in public organizations generates new routine capability in advancing human and machine practices.

Similarly, the chapter by Lindgren, Åkesson, Thomsen, and Toll illustrates how two Swedish municipalities have organized for automation and RPA initiatives. The study shows how politically driven policies on process automation become a driving force that challenge traditional ways of organizing work and IT in Swedish local governments.

Similarly, Bygstad, Övrelid, and Williams address the challenges of managing lightweight and heavyweight IT innovation. They propose a governance framework for two-speed innovation which is illustrated in three empirical cases. The study illustrates that RPA clearly offers new opportunities for public sector organizations, but that these opportunities are accompanied by a set of challenges that, according to the authors, can be balanced with the framework of two-speed innovation.

Lastly, and as mentioned in the introduction, RPA was developed for business purposes and has been deployed in the private sector for some time. Drawing on his research on RPA use in the private sector, Asatiani offers advice to public sector organizations embarking in automation and RPA initiatives.

Together, these last four chapters provide important insights on the need for public organizations to understand that automation technologies can result in more substantial organizational changes than first anticipated. An important lesson learned is that seemingly simple applications such as RPA requires that the organization invests in new skillsets and competencies, as well as new organizational structures.

3 Future of Public Service Automation

When combining the research efforts presented in the chapters, we see that service automation is in fashion for public organizations. Using the words of Wang (2010), RPA has been presented as the "hottest IT" for public organizations, resulting in an array of initiatives to automate parts of public service delivery. So far, research on this topic is limited and there is great potential for enhancing the understanding of the topic of public service automation.

The chapters focus mainly on public organizations' use of RPA. However, our ambition is to understand service automation in a wider sense and to consider RPA as a useful empirical example of service automation in public organizations. At present, RPA is in fashion but will most likely be considered as a commonplace or outdated technology in a foreseeable future (like other hyped technologies before). In contrast, automation of work and services, as a more general concept, is likely to remain a relevant and evergreen topic for public organizations. With upcoming advancements in AI-based applications and services, we furthermore anticipate an increase in automation of work and service delivery in the public sector. The lessons learned from these initial applications of RPA are, therefore, important for understanding the future implications and challenges of public service automation, and call for further research:

- Practitioners and suppliers of service automation have high hopes that automation will lead to lower costs, more efficient processes, fewer errors, time savings, improved service quality of citizen services, and higher work quality for civil servants. But so far, there is still little evidence of these effects of automation of public service; future research has an important task to evaluate these effects over time.
- There is a need for more research on how specific automation technologies and solutions are developed and on what grounds specific work practices are chosen and re-designed for automation. Furthermore, we see a need for more studies on the co-creation of new automated work practices. Considering that automation is co-created by a large set of actors, design thinking can be a fruitful perspective for creating a deeper understanding of how automation solutions are negotiated and created, based on both technical- and social dimensions of work.
- The social consequences of increased public service automation are not well understood. Increased automation will change the work situation for many professionals in the public sector and affect interaction and communication patterns in and between public organizations, and in relation to citizens. Similarly, the citizen perspective as well as the policy perspective is still under-researched. So far, little is known about how service automation is perceived by citizens and how automation affects policy making. For example, future research could shed light on how citizens' trust in authorities is affected by increasingly automatic decisions.
- There are several policy areas within the public sector yet to study. The applications covered in the book are taken from social work and human resource management. Therefore, we expect to see future research about cases of service automation in other policy areas, as well as on how to scale automation within and between policy areas.
- It can be highly valuable to draw lessons from private organizations' use of automation and transfer these lessons to the public sector. At the same time, it is also important to recognize that there are significant differences between the sectors. How the specifics of public sector organizations affect their opportunities to automate work and service delivery must be investigated further. It is also important to investigate the influence of commercial actors on the design of public organizations and processes, e.g. the role played of technology suppliers and consultants in reshaping the public sector.

- Last, but not least, there is a need for a critical perspective on automation in the public sector; what is possible versus desirable? What are the risks of the digitally automated society? There are, for example, issues of transparency and traceability in automated processes and algorithm-based decision-making that require further research.

In this book, we contribute to the understanding of service automation in the public sector with studies from three perspectives: conceptualization, applications, and implementation challenges. The results reveal several questions for future research, as outlined above. We hope that this book will inspire to new and exciting research initiatives, perhaps with the aim of answering some of these questions.

References

Bhatnagar, N. (2020). Role of robotic process automation in pharmaceutical industries. In A. Hassanien, A. Azar, T. Gaber, R. Bhatnagar, & M. F. Tolba (Eds.), *The International Conference on Advanced Machine Learning Technologies and Applications (AMLTA2019). AMLTA 2019. Advances in Intelligent Systems and Computing* (Vol. 921, pp. 497–504). Springer.

Bygstad, B. (2012). Generative innovation: A comparison of lightweight and heavyweight IT. *Journal of Information Technology, 32*, 180–193.

Lacity, M., & Willcocks, L. (2021). Becoming strategic with intelligent automation. *MIS Quarterly Executive, 10*(2), 1–14.

Lindgren, I., Madsen, C. Ø., Hofmann, S., & Melin, U. (2019). Close encounters of the digital kind: A research agenda for the digitalization of public services. *Government Information Quarterly, 36*(3), 427–436.

Lindgren, I., Toll, D., & Melin, U. (2021). Automation as a driver of digital transformation in local government: Exploring stakeholder views on an automation initiative in a Swedish municipality. In *DG.O2021: The 22nd Annual International Conference on Digital Government Research (DG.O'21)*, June 09–11, 2021, Omaha, NE, USA. ACM, New York, NY, USA, 10 pages. https://doi.org/10.1145/3463677.3463685

Ranerup, A., & Henriksen, H. Z. (2019). Value positions viewed through the lens of automated decision-making: The case of social services. *Government Information Quarterly, 36*(4), 101377.

Ranerup, A., & Henriksen, H. Z. (2020). Digital discretion: Unpacking human and technological agency in automated decision making in Sweden's Social Services. *Social Science Computer Review*, 1–17.

Suri, V. K., Elia, M., & van Hillegersberg, J. (2017). Software bots - the next frontier for shared services and functional excellence. In I. Oshri, J. Kotlarsky, & L. Willcocks (Eds.), *Global sourcing of digital services: Micro and macro perspectives. Global Sourcing 2017. Lecture Notes in Business Information Processing* (Vol. 306, pp. 81–94). Springer.

Toll, D., Lindgren, I., & Melin, U. (2021). Process automation as enabler of prioritized values in local government – A stakeholder analysis. In H. J. Scholl et al. (Eds.), *Electronic Government. Proceedings of the 20th IFIP WG 8.5 International Conference, EGOV 2021, Granada, Spain, September 7-9, 2021* (pp. 288–300). Springer, LNCS 12850.

Wang, P. (2010). Chasing the hottest IT: Effects of information technology fashion on organizations. *MIS Quarterly, 34*(1), 63–85.

Wihlborg, E., Larsson, H., & Hedström, K. (2016). The computer says no! – A case study on automated decision-making in public authorities. In *Proceedings from 49th Hawaii International Conference on Systems Sciences*.

Wirtz, B. W., & Müller, W. M. (2019). An integrated artificial intelligence framework for public management. *Public Management Review, 21*(7), 1076–1100.

Part II
Conceptualization of Public Sector Service Automation

The Subject Matter of Process Automation Practices: Through the Lenses of Research Questions

Göran Goldkuhl

1 Introduction

In its inception, the computer was considered and used as an automatic machine. This means that it was self-operating by control of software in transforming input data to output data. The computer was used for numeric and later administrative applications (Mahoney, 2005). The roles of humans were to feed the machine with data and after computing interpret and use its produced results. Later in history, a radical change occurred through the introduction of the micro-computer and its graphical user interface (ibid.). The computer was transformed from previously the main role of an automaton to the role of an interactive tool (Ehn & Kyng, 1984). This tool view has later been emphasized through the development and integration of communication technologies including the Internet revolution. Interactivity has not taken away automatic functionality of computers, but only made this less visible and obvious. Computers operate, as controlled by software, and transform input to output. These computing steps are just smaller in an interactive mode than in large-scale computing where the automatic functions are more obvious.

Already at the outset, automation was closely associated with computers and information technology. It has been used as a buzzword on several occasions during history, as in automation and robotics in industrial engineering (Mahoney, 2005) but also in office automation (Olson & Lucas, 1982). It would be possible to talk about different automation waves during the history of computing. A recent wave is that of robotic process automation (RPA); e.g., Lacity and Willcocks (2016). The main idea is to reduce the human involvement in administrative processes; to have longer chains of automatic computing without any humans maneuvering the IT artifact. The

G. Goldkuhl (✉)
Information Systems and Digitalization, Department of Management and Engineering, Linköping University, Linköping, Sweden
e-mail: goran.goldkuhl@liu.se

G. Juell-Skielse et al. (eds.), *Service Automation in the Public Sector*, Progress in IS,
https://doi.org/10.1007/978-3-030-92644-1_2

role of the IT artifact returns to that of an automaton instead of being an interactive tool. In this new type of technology, the RPA-based software will be the interactor with other IT artifacts' user interfaces (Lacity & Willcocks, 2016). This and other types of technology have found their ways to the public sector where they are used for automating back-office processes such as case handling and routine decision-making (Wihlborg et al., 2016; Ranerup & Henriksen, 2019; Houy et al., 2019). Process automation in the public sector can be conducted with the aid of specific RPA software (Lacity & Willcocks, 2016) or other types of software (Houy et al., 2019).

As indicated above, process automation in the public sector is both an old and new phenomenon. There is an obvious need for a conceptually clear view of this phenomenon in research (Wihlborg et al., 2016; Ranerup & Henriksen, 2019, 2020; Houy et al., 2019). This chapter is based on an ongoing meta-study of several research projects on process automation in the public sector. The main purpose is to elaborate on the subject matter of process automation (PA). To clarify ontologically, the phenomenon of process automation can be made in different ways. The route taken in this chapter is an indirect one. The subject matter of process automation is studied mainly through the eyes of other researchers that conduct PA research. The research interest in this chapter translates into different inquiry questions: How is process automation framed and conceptualized in the research questions of different research projects? What kinds of phenomena appear as salient in the problem formulations that guide the research? The chapter contributes to an enhanced understanding of how to conceptualize process automation in the public sector.

2 Research Approach

As stated, the basic inquiry idea delineated for the chapter is how to view and conceptualize the research subject matter of process automation; see Fig. 1 for an overview. The focus is on pre-conceptualizations made in research questions (RQs). Through the study of five research projects on process automation, a comprehensive view of its subject matter appears. This emergent view should have greater coverage than each singular view from the research projects, respectively.

The source material for the meta-study consists of:

- Research project descriptions and other related material
- Interviews with researchers

The chapter presents a review of such pre-conceptualizations on process automation. The main purpose is not a comparative review of studied research projects. I use RQs of five research projects to obtain a richer view of the subject matter of process automation. This analysis of five research projects produces inevitably a material that makes a comparison of those projects possible.

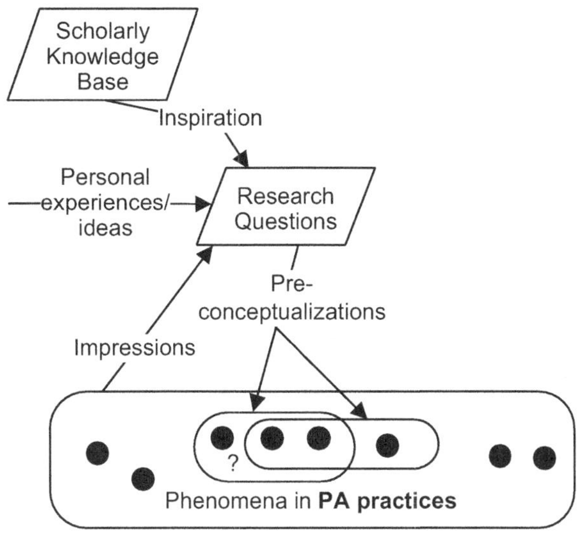

Fig. 1 Research overview

The review of pre-conceptualizations is made through linguistic and ontological analyses of basic concepts. This conceptual analysis is based on the "linguistic turn" (e.g., Alvesson & Kärreman, 2000; Goldkuhl, 2002), the "practice turn" (e.g., Schatzki, 2001; Reckwitz, 2002; Goldkuhl & Röstlinger, 2006; Nicolini, 2012) and processual and relational thinking (Abbott, 1992; Emirbayer, 1997; Goldkuhl, 2005).

Initially, I investigate the concept of process automation from a linguistic perspective as a basis for further analysis. Then, research questions (from the studied research projects) are investigated in a two-step procedure. First, the types of practices that are presumed in the RQs are articulated. Second, different phenomena mentioned and presumed in the RQs are identified and ontologically determined following the ontology of socio-instrumental practice theory (Goldkuhl, 2002, 2019; Goldkuhl & Röstlinger, 2006).

The outcome of this meta-inquiry of different research projects is a suggested and synthesized description of process automation practices. This is made in the last step of this inquiry: Different practice reconstructions and the conceptual analysis of phenomena from RQs are synthesized in an ontological map of process automation practices as a conclusion of this chapter.

3 A Linguistic Analysis: What Does "Automation" Mean?

As a preparatory step, before I embark on a review of the studied research projects, I look into possible meanings of the main concept "process automation." I follow the lines of the linguistic turn, which means a turn to an enhanced focus and understanding of the role that language plays in scientific development (e.g., Alvesson & Kärreman, 2000). "Process automation" is composed of two words/concepts. "Process" is what is automated, and in this context, it is processes of public administration, and more specifically it will often be case handling processes that may involve automated decision-making. In this initial linguistic analysis, I will not go into any detail concerning such processes. This is done in the following analysis of the research projects below. Instead, my focus is on "automation" and other related words/concepts. When I use this "word/concept" constellation, it is an indication of the type of analysis that I pursue, i.e., a mix of linguistic and conceptual analyses. A linguistic analysis means an analysis of the types of words used for denoting categorized phenomena, as processes (normally expressed as verbs), agents and entities (normally expressed as nouns), and properties (normally expressed as adjectives); see Goldkuhl (2002). The conceptual analysis has of course some influence from ontological reflections, otherwise, it should be conducted in a vacuum free of real phenomena. "Automation" is the keyword to reflect on here but there are other related words such as "automatic," "automaton," "automate," and "automated." A summary overview of my linguistic analysis can be found in Fig. 2. Concepts are in bold, a linguistic determination in italics followed by a conceptual clarification.

First, the concept of automation builds on the existence of "automatons" as technical artifacts with agency power. Such entities/agents must have properties of being "automatic" (self-operating functions). These two words/concepts are foundational for analyzing concepts related to automation (see Fig. 2). Their etymological roots are in the Greek word "automatos" with the meaning of self-acting (www.etymonline.com).

The word "automation" is not completely unequivocally. It is a noun and as such it has its roots in the verb "automate." To automate is to develop and arrange something with automatic functions. However, "automation" might also be interpreted as practices being automated, thus the result of automating. This duality of meanings is actually established in dictionary. The Merriam-Webster dictionary gives, for the entry "automation," these two definitions: (1) "the technique of making an apparatus, a process, or a system operate automatically" and (2) "the state of being operated automatically" (www.merriam-webster.com). How come that this duality of meanings has been established in language? The performed linguistic analysis suggests a possible explanation as described in Fig. 2. The first meaning (automation as development) is obtained as a direct nominalization of the verb "automate." Both words are process-words denoting the same type of phenomenon; "automate" is a verb, and "automation" is a noun, which is a nominalized process-word. This is depicted on the left side of Fig. 2.

On the right side of Fig. 2, a linguistic trajectory in two steps is shown how "automate" is transformed to "automation" with the meaning of use of automatons. The verb "automate" can give rise to the word "automated," which is a participle in grammatical terms (step one). Such a word-form has an origin in a verb but is used as an attribute (Eastwood, 1994). "Automated" is thus an attribute of something with the meaning of being automated (the result of automating). This attribute in form of a

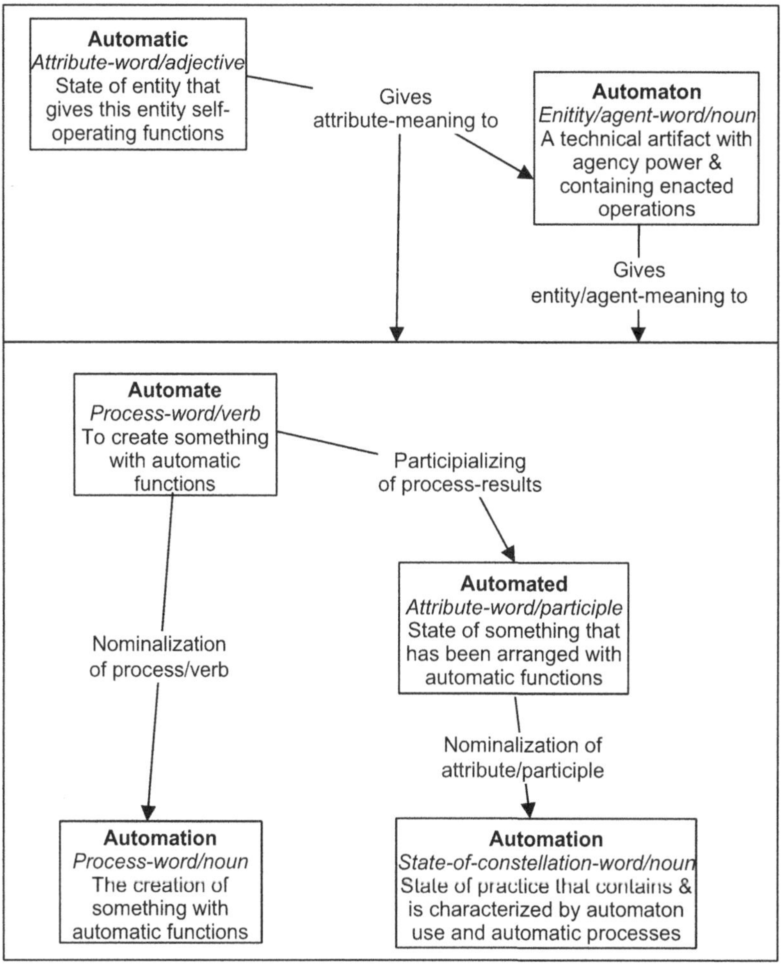

Fig. 2 A linguistic chart

participle is then, in step two, nominalized to the word "automation." This derived word is an "attributive noun" denoting the state of the phenomena being automated, which may cover not only automatons but also the social use context of those artifacts. "Automation" denotes in this second meaning the phenomena that result from the process of automating.

Homonyms, like these, can of course be problematic in scientific development. How can we be sure, in every instance of the used homonymic word, which meaning is intended? The conclusion from this linguistic analysis is that we need to be cautious when using the word "process automation." My suggestion is that we can use this word with the meaning of a scientific domain of interest which may cover both (1) the practices of developing/automating and (2) the practices of being automated. For the sake of being more exact, I suggest that (1) can be labeled "automating practice"; here using the gerund form (Eastwood, 1994) to emphasize its processual character. For (2), the label "automated practice" is suggested, where the past participle form emphasizes the phenomena as results from development work. I use this proposed terminology in the inquiry below (Sects. 4–6).

4 A Practice Analysis: Identification of Presumed Practices

4.1 A Practice-Theory View

The research questions are investigated from a practice-theory view. This is founded on the so-called *practice turn* in research (e.g., Schatzki, 2001; Nicolini, 2012). A practice-theory view emphasizes the *performance* of activities and actions. A practice is considered a meaningful and generative *constellation of agents, activities, and objects* (ibid., and Goldkuhl, 2002, 2005, 2019; Goldkuhl & Röstlinger, 2006; Reckwitz, 2002). A practice implies doing. There are agents that conduct activities, actions, and operations of diverse kinds. Practice consists of *symbolic* and *physical objects*, sometimes mixtures of such objects. Symbolic objects are created for communication, and these objects can be talk (temporary/elusive oral objects) or written/recorded (manifested more or less permanently in some way). Activities can be conducted by *humans* (individually or collectively) or by *artificial agents* as, e.g., digital artifacts. Practices can be performed in *organizational* and *professional settings*. This means that organizations (as institutional actors) conduct actions. However, an organization cannot by itself conduct actions. There must exist *organizational representatives*, such as human or artificial agents that conduct organizational actions in name of that organization.

The performance of activities within a practice is governed and guided by *shared practical knowledge* (Schatzki, 2001). Such practical knowledge comprises knowledge about typical objects, processes/routines, roles, values, rules, and the language used within that practice (Goldkuhl & Röstlinger, 2006). There exists an overall purpose, meaning, and logic that holds a practice together. Shared knowledge and language imply *dispositions* for practice performance and make the practice a *coherent social unit*. The practice purpose determines which actions are adequate within that practice; what actions count as enactments of the practice.

Practice theory relies on *processual* and *relational thinking*. There exist always *meaningful situations* in a practice. These are constellations of agents, activities, and objects. There are never objects that exist in isolation, totally unrelated (Dewey, 1938). Elements of practices are related to each other in meaningful and dynamic ways (Abbott, 1992; Emirbayer, 1997).

4.2 Presumed Practices in Research Questions

The five studied research projects are governed by stated research questions. These RQs are analyzed in Sect. 5. Each research question contains a research focus on some phenomena. These phenomena are presumed to exist and occur in some kind of practice. The review of RQs has thus resulted in an identified set of practices. This set of practices is an outcome of my review, but I present it here before the review account (Sect. 5) in order to easily make references to the different kinds of practices when presenting the projects' RQs. The set of practices are depicted in Fig. 3.

Eight practice types are visualized in Fig. 3. First of all, there is an automated case handling practice (P5). This is an operational practice of continual and permanent character. Such a practice is a "new version" of a previously non-automated version (P1). When I use the term "non-automated," this does not mean "non-digitalized." I presume that there exists some digital support already in the previous version of the case handling practice. However, such previous digital support is not based on the ambitions of automating case handling. Six other practices appear also in the practice model. The construction and generation of specific PA software (which is labeled "constructed artifact") is made in a practice called development (P3). The construction of a PA artifact implies the use of digital technologies and tools. The selection of PA projects is made in a practice called IT governance (P2). The constructed artifact needs to be implemented in the case handling practice. This is done through an implementation practice (P4) that transforms the nonautomated case handling practice (P1) into an automated one (P5). Adapted instrumental support (as methods and IT tools) may be needed during development and implementation. The provision of such instrumental support is located in the practice called IT governance. This

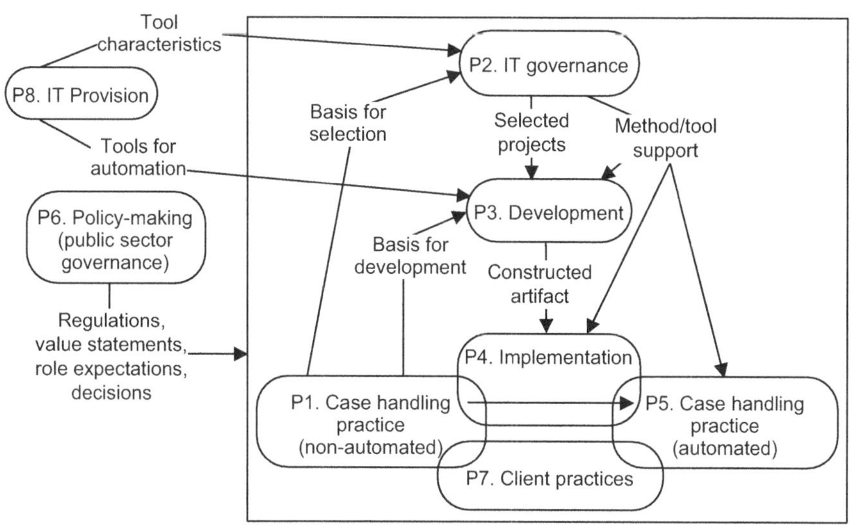

Fig. 3 A set of practices related to process automation

strategic practice (P2) is directly concerned with the governance of digitalized practices. There exist also other strategic practices in public organizations. Such organizations are governed by overall public policy-making practices (P6). These practices include policy-makers such as politicians and public agency managers. The result of such policy-making comprise regulations, value statements, role expectations, and strategic decisions concerning public administration. I here take a broad and inclusive interpretation of public policy following Lascoumes and Le Gales (2007). Policy-making should not be seen as one single practice, rather as a set of dispersed practices.

Besides these different practices of public administration (P1–P6), I have also included, in the practice model, client practices (P7). Clients are citizens in different roles in relation to public administration dependent on what kind of authority or service is exercised; for example, social service clients related to municipal social assistance. In Fig. 3, there is an overlap between public case handling (P1/P5) and client practices (P7). Clients can be seen as being "visiting constituents" of case handling since (1) they produce applications and other material for the handling of cases, and (2) they are also the key recipients of decisions from the exercise of public authority in the handling of the cases.

I have also included external practices of IT provision (P8) in the model. These are practices of IT vendors that provide general tools for automation to be used in public sector development. Characteristics of such digital tools are assessed in IT governance (P2) in the choice of what digital technologies to use for automation. The provision of adapted instrumental support (methods and tools) is, as said located in the practice of IT governance (P2).

The three practices of IT governance, development, and implementation (P2, P3, P4) are seen as *automating* practices, and the resulting case handling practice (P5) is an *automated* practice (following the developed terminology in Sect. 3). The automating practices of development and implementation are, as projects, temporary practices. IT governance (P2), as practices of strategic character, may be of more permanent character, but can also appear as temporary practices dependent on the kinds of IT governance issues to address.

As can be seen from the practice model (Fig. 3), the different practices are related to each other. All possible relationships are not covered in this overview model; many others exist. A professional work-practice has intended outcomes aimed for one or more other practices. This means that a practice has through its intended outcomes other practices as *target practices*. For example, the case handling practices (P1/P5) has client practices (P7) as their target practices. A practice can have several subsequent practices as its target practices. For example, IT governance (P2) produces outcomes that have governance functions for development (P3), implementation (P4), and automated case handling (P5). IT governance is also concerned with nonautomated practices (P1), through analysis of pre-conditions and automation options. This means that activities within a practice can pertain to another practice that is not a target practice. A practice that is studied or dealt with in some way in a focused practice can be called a *referent practice*, as being an object for that focused practice.

In my analysis below, I use the codes for the practices (P1–P8), which are visualized in Fig. 3. I have tried to maintain the terms from the projects but in some cases, I have made some re-formulations based on my linguistic and ontological analyses.

5 A Review of Research Questions in Selected Research Projects

This meta-study contains investigations of five Swedish research projects on process automation. My descriptions below are based on project descriptions and interviews. When needed, I have made translations from Swedish. Projects and their academic hosts are:

- From form to robot? Automated case handling in Swedish municipalities (Linköping); see Lindgren (2020) and Lindgren et al. (2021).
- Digital discretion in social services: The case of automated decision-making (Göteborg); see Ranerup and Henriksen (2019, 2020).
- The computer says no! A study of public sector legitimacy and the trust of citizens when e-government emerges (Örebro); see Wihlborg et al. (2016) and Andersson et al. (2018).
- AI in service of the bureaucracy—a changed digital work environment when robot colleagues become part of everyday life (Halmstad); see Åkesson and Thomsen (2020).
- A case study of RPA implementation in public organizations (Stockholm); see Güner et al. (2020).

The analysis of each project consists of two parts. First, an identification and analysis of presumed practices in RQs. Second, a practice-ontological determination of mentioned phenomena in RQs (Tables 1, 2, 3, 4 and 5).

5.1 *The Linköping Project*

The Linköping project has the following three stated RQs in their project application authored by project leader Ida Lindgren:

1. What meanings of development of automated case handling in Swedish municipalities; i.e., how should process automation be interpreted and understood?
2. What are the consequences of automated case handling in Swedish municipalities?
3. What pre-conditions must exist in order to review and assess if and to what extent a specific case handling process should be automated?

Table 1 Phenomena in RQs of Linköping project

Phenomenon	Ontological characterization
Automated case handling (ACH)	Digital activities that are performed by a digital artifact and especially designed for automatic performance (with aminimum of human interactivity).
Development of ACH	*Activities performed by humans*, with instrumental aid of general and/or specific software, for generation of a digital artifact aimed for ACH.
Implementation of ACH	*Activities performed by humans* in order to technically and organizationally implement an ACH artifact. These activities aim at (1) assuring that the digital artifact is technically interoperable in the current digital landscape and (2) transforming the nonautomated case handling practice to an automated one through informing, training, and routine rearranging.
ACH consequence	Consequences that appear during the use of a digital ACH artifact. Consequences are intentional or unintentional *influences on the practice context* of a digital ACH artifact and possibly also more widely *on other related practices*.
Stakeholders	*Group of actors* with a specific role that is related to automated case handling.
Stakeholders' work situation	Stakeholders' (1) *performance of their work tasks* framed by (2) work demands and other *environmental conditions of physical and symbolic character* and (3) their consequential *work experiences*.
Pre-conditions for development of ACH	Different *properties of a case handling practice* that may have significance for automating such a practice.
Automation candidate	*Idea* generated by one or several stakeholders about a potential automation effort and a *description* of this idea with the purpose to influence change planning in the organization.
Review of automation candidates	*Activities performed by humans* to assess automation candidates and make decisions about development of digital ACH artifacts.
Analysis and method support (for automation review)	*Guidelines* for the automation review activities. Such prescriptions may appear as *intersubjective and practical knowledge* among decision makers and as *documented prescriptions*.

Overall, this project uses the concept of *automated case handling*. In RQ1, the word *development* appears. This indicates an interest in a development practice (P3), however, as stated in a complementary description, this also includes *implementation* (P4) and other change issues. From clarifications in writing and interview, this RQ-interest is not limited to developmental and change issues. There is a search for knowledge about the purport of automated case handling as such, i.e., what characterizes automation in a public administrative practice (P5). This would also presume knowledge about nonautomated practices (P1) as contrasting knowledge. To know what is particular about automation, one needs to understand how it is different from non-automation. RQ2 is clearly stated as an interest in an automated practice (P5) and is particularly concerned with *consequences* from such use. It appears from a project description an interest in consequences on *stakeholders* and their *work situations*.

RQ3 is a more complex research question. It has a focus on processes of *reviewing automation candidates* (P2). This RQ also contains references to *pre-conditions* for reviewing and deciding on automation efforts. Such pre-conditions are, at least partially, significant properties of existing administrative processes (P1). The research project has an explicit ambition to develop and contribute an *analysis and method support* for reviewing automation candidates and further such initiatives to development.

In Table 1, I have listed salient phenomena that appear in RQs and in complementary clarifications of RQs. These different phenomena have been ontologically analyzed following practice theory; see the second column of Table 1. This analysis of phenomena mentioned in the Linköping project (as well as the other projects below) has been performed based on the ontology of socio-instrumental practice theory (Goldkuhl, 2002, 2005, 2019; Goldkuhl & Röstlinger, 2006).

5.2 The Göteborg Project

The Göteborg project has the following two stated RQs in their project application authored by project leader Agneta Ranerup:

1. What value positions do involved actors (politicians, project leaders, professionals) hope to strengthen through automated decision-making related to social assistance?
2. What are the more immediate and long-term effects of digitalization and automated decision-making on civil servants' discretion and law-proof decision-making from a social work profession perspective?

This project has a focus on *digital discretion* and *automated decision-making* in (the public sector area of) *social services*. This project uses thus the concept of decision-making as the key concept for automation. In both RQ1 and RQ2, the focus is on automated decision-making, which corresponds to practice P5 where the covering notion of "case handling" is used. In RQ1, automated decision-making is related to different *value positions* that are expressed and held by different *stakeholders* (actors). This means that such value positions may emerge from different practices. Politicians are mentioned as one such stakeholder. This means that such value positions emerge from a *policy-making practice* (P6). Another mentioned stakeholder type is project leader. The active work of a project leader should be performed in the practices of *development* (P3) and *implementation* (P4). A third stakeholder group is "professionals," which I equate with *case handlers* that are the actors of the focused practices of P1 and P5. In RQ2, the concept of civil servant is used, which I also equate with case handler. The research focus in RQ2 is concerned with effects of automated decision-making on case handlers' discretion and how automation affects *law-proof decision-making* (legal security). This explicit reference to laws means a reference to legislating as one type of policy-making practice (P6). In RQ2, a *social work perspective* is explicitly mentioned. This means

Table 2 Phenomena in RQs of Göteborg project

Phenomenon	Ontological characterization
Automated decision-making (ADM)	Digital activities that are performed by a digital artifact and especially designed for automatic performance (with a minimum of human interactivity) and which include the generation of official decisions.
Social assistance	*Activities* conducted by the public sector (mainly social workers) to support clients in need with practical, administrative, and economic matters based on social service regulations.
Social work profession	*Know-how* and *pursuance* concerning social assistance are *shared* among those professionals within that profession.
Value position	*Position* held/*expression* made by stakeholders about *desirable states*. An expressed value position can be *intersubjectively* and *institutionally* established.
Discretion	*Action and decision space* in public case handling *as perceived* and *applied* by case handlers due to norm expectations expressed in formal and informal statements.
Law-proof decision-making	*Activities* of public decision-making that are conducted in accordance with *stated laws*.

professional knowledge and exercised competencies held by this type of case handlers that operate in the practice of P1 and P5.

In Table 2, I have listed salient phenomena that appear in RQs and in complementary clarifications of RQs. These different phenomena have been ontologically analyzed; see the second column of Table 2. In the RQ1, the term "actors" appears and this is considered as equivalent with "stakeholders" in Table 1. "Effects of automated decision-making" (RQ2) is considered as equivalent with "ACH consequence" in Table 1.

5.3 The Örebro Project

The Örebro project has the following four stated RQs in their project application authored by project leader Karin Hedström:

1. When case handling and decision-making are automated, how does this influence case handlers' discretion and possibilities to exercise competence and accountability?
2. When case handling and decision-making are automated, how and why are organizing and arrangements of accountability relations changed?
3. (a) When case handling and decision-making are automated, what interpretations of client diversity made by case handlers are enabled vs. inhibited? (b) How do case handlers explain this and what other explanations can be found?
4. How can analyses and experiences from automated case handling and decision-making contribute to the understanding of public sector legitimacy?

The RQs use the comprehensive concept of *automated case handling and decision-making*. All four RQs are concerned with such automated practice (P5). RQ1 relates automation to important case handling phenomena such as case handler *discretion*, *accountability*, and their *exercise of professional competence*. In RQ2, the focus is on accountability and how such *organizational arrangements* are made due to automation. RQ3 consists actually of two related questions (RQ3a, RQ3b). These questions are concerned with case handlers' interpretations of client properties and circumstances ("client diversity"). This includes thus an explicit focus on clients and client practices (P7). RQ3 is also concerned with how automation *influences* (in terms of *enabling* and *inhibiting*) such case handler interpretations of client properties. RQ3b includes an explicit interpretive stance on how case handlers interpret and *explain* such influences. All three mentioned research questions (RQ1–3) build on some comparison between an automated practice (P5) and a nonautomated practice (P1).

RQ4 is not formulated with direct reference to empirical matters. It is instead oriented toward the use of data analysis from the project for the development of public sector legitimacy understanding. In the context of process automation, I have interpreted this as a knowledge interest in the *legitimacy of public sector* case handling and decision-making. This means assessments made by citizens/clients concerning such case handling. This pertains mainly to the client practices (P7) and the clients' interpretations and assessment of P1/P5 and probably also to some extent to policy-making/governance including legislation (P6).

In Table 3, I have listed salient phenomena that appear in RQs and in complementary clarifications of RQs. These different phenomena have been ontologically analyzed; see second column of Table 3. "Automated case handling and decision-

Table 3 Phenomena in RQs of Örebro project

Phenomenon	Ontological characterization
Accountability	*Intersubjectively ascribed role attribute* of case handlers concerning their responsibility for decisions and other actions. Role ascription is based on regulations and administrative praxis, and implies role expectations on the case handlers.
Organizing and arranging accountability relationships	*Activities* of key stakeholders (mainly *symbolic actions*) to clarify and state case handler accountability.
Exercise competence	Case handler *activity* when evolved *competencies* are *utilized* and *applied.*
Interpretation of client diversity	*Subjective interpretations* made/held by case handlers about specific client properties and circumstances that are considered relevant for handling the case.
Influence (enablers, inhibitors)	*Efficacies* in case handling that imply and comprise *pre-conditions* as enabling or inhibiting *actions* and *results/effects* of actions.
Case handler explanations	*Subjective interpretations* made/held by case handlers about effects of automation and expressed in *stated explanations.*
Public sector legitimacy	Citizens'/clients' *interpretations, judgements,* and degree of *acceptance* about the rightness of public sector exercise of power.

making" is equivalent to what is stated in Tables 1 and 2. "Discretion" is equivalent to what is stated in Table 2.

5.4 The Halmstad Project

The Halmstad project has two stated RQs in their project application authored by project leader Maria Åkesson. The English versions of these two RQs are taken from Åkesson and Thomsen (2020). Besides these, I have added one more RQ (3) that appears from the description of project activities.

1. What are the implications of RPA and AI for the digital working environment?
2. How can RPA and AI be implemented to support good bureaucracy and good digital working environment?
3. What kind of digitalized guidelines can support development and implementation of RPA and AI to reach good bureaucracy and good digital working environment?

The different RQs do not contain explicit references to case handling activities (P1, P5), but the use of "RPA" and the complementary project descriptions make this obvious. RQ1 and RQ2 focus *digital working environment* after automation (P5), which presumes some comparison with nonautomated case handling (P1). In RQ2, there is also reference to case handling as a bureaucratic process. This presumes the application of different *normative* and *regulative principles* in case handling. Such principles are settled in public sector governance (P6) and exist as *applied knowledge* in administrative processes (P1) and will thus historically emerge from such established practices (P1 → P4 → P5). RQ2 uses explicitly the concept of *implementation* (P4). It is here presumed that this also covers "*development*" (P3), which is clear from RQ3.

The three RQs use the terms RPA and AI to refer to these specific *digital technologies*. This means that this research project is more specific concerning what digital technologies are to be applied in process automation. Therefore, I also add an explicit reference to the external practice of digital tool provision (P8).

RQ3 focuses on *digitalized guidelines* for development (P3) and implementation (P4) of automated case handling (P5). Following the structure of practices (Fig. 3), the creation of such guidelines is found in the IT governance practice (P2).

In Table 4, I have listed salient phenomena that appear in RQs and in complementary clarifications of RQs. These different phenomena have been ontologically analyzed; see the second column of Table 4. This ontological analysis relies on previous analyses, especially the Linköping project (Table 1). I refer to this table for "automated case handling," "development" and "implementation." These terms are not explicitly described in Table 4. Important phenomenon in the Halmstad project is "implications for the digital working environment of ACH" (RQ1). ACH implications are equated with ACH consequences as described in Linköping Table 1.

Table 4 Phenomena in RQs of Halmstad project

Phenomenon	Ontological characterization
Process automation technology	Basic *technological principles* and *functionalities* (such as Robotic Process Automation or Artificial Intelligence/machine learning) that are used in *digital tools* for automation of administrative case handling.
Bureaucratic principles in ACH	*Normative*, *regulative*, and other *cognitive principles* that pertain to and are *shared* and *applied* in ACH both by *human* and *digital agents*. Such principles can emerge from explicit processes of formulation in public sector governance or through continuous application in administrative processes.
Digitalized guidelines for ACH development & implementation	*Guidelines* for automation development and implementation activities. Such guidelines appear first as *documented prescriptions* in *digital tools* and may also appear as *intersubjective and practical knowledge* among stakeholders working with ACH creation.

Working environment of ACH is equated with stakeholders' work situation as described in Linköping Table 1.

5.5 *The Stockholm Project*

The Stockholm project has formulated several purposes for their first case study. These purposes have been transformed, by me and checked by project leader Gustaf Juell-Skielse, into three RQs:

1. What goals govern the adoption of RPA technology for the creation of process automation and how are automation changes managed?
2. What benefits, planned and other changes occur in the use of robotic process automation?
3. What kind of guidelines are needed for the development, implementation, and use of robotic process automation?

The RQs have a focus on *change* and related concepts such as *goals* and *benefits*. In my ontological analysis, I make a distinction of "change" to mean (1) *changing as process* and (2) change as an *altered circumstance in a practice*. Change process pertains to the practices of development (P3) and implementation (P4). In RQ1, there is a reference to "management of change" that may, besides P3 and P4, also include reference to continual adaptation and adjustment in an automated case handling practice (P5). Change as an altered practice circumstance covers both (2a) implemented objects such as a developed automation artifact (a *direct result* of change efforts) and (2b) *effects from the use* of such an artifact; confer this distinction of result vs. effect in Ryle (1949) and Goldkuhl (2005). Effects may appear in the automated case handling practice (P5) as well as effects on/in client practices (P7). The concept of change covers thus both implemented new objects and altered activities/routines in a practice and effects that occur from use. The concept of effect, in this ontological analysis of the Stockholm project, corresponds to the concept consequence in the Linköping project (Sect. 5.1), effect in the Göteborg project (Sect. 5.2), and implication in the Halmstad project (Sect. 5.4).

RQ1 focuses on the adoption of RPA technology for the creation of automated case handling. Creation seems to cover *development* (P3) and *implementation* (P4) practices. The reference to the management of automation changes can comprise, besides P3 and P4 also the automated practice as such (P5). There may be a continuous adaptation of altered circumstances that need to be dealt with. Goals that govern the creation of process automation can be articulated during development, but can have an origin in other practices, such as existing case handling (P1) and governance practices (P2, P6). The reference to RPA as specific technology, in this and the other RQs, opens for a link to IT provision (P8). RQ2 refers to the automated case handling practice (P5) where benefits and other changes occur as distinct from previous nonautomated case handling practice (P1). The planning of benefits/changes will mainly be done during development (P3) but can also have an origin in IT governance (P2). In RQ3, *guidelines* are focused. These guidelines aim to support creation, i.e., development (P3) and implementation (P4), of automated case handling (P5). Guidelines pertain also directly to the *use of automation* (P5). Following the practice map (Fig. 3), guidelines are created in the IT governance practice (P2).

In Table 5, I have listed salient phenomena that appear in RQs and in complementary clarifications of RQs. These different phenomena have been ontologically analyzed; see the second column of Table 4. The ontological analysis of the

Table 5 Phenomena in RQs of Stockholm project

Phenomenon	Ontological characterization
Actual and planned ACH changes	New or altered *situations* or *circumstances* in case handling practice or influenced client practices (that *differ* from previous version of that practice). An actual change can be a *direct result* of developmental work or *occur* as an *effect* of such a result. An actual change can be *planned* and *intentional* or occur as *unintentional/unanticipated*. Changes in ACH are mainly planned during ACH development.
Goals for ACH	Goals are *held* and *shared* among stakeholders and can be expressed as *normative statements* about situations in practices that are found *valuable*. Goals can *govern* developmental work and the performance of case handling practices.
Benefits in/of ACH	*Situations* or *circumstances* in case handling practice or influenced client practices that are *considered valuable* by and for some stakeholders.
Guidelines for ACH development, implementation, and use	*Guidelines* for automation development and implementation activities and automation use. Such prescriptions may appear as *intersubjective* and *practical knowledge* among stakeholders and as *documented prescriptions*.

Stockholm project is the last one and it relies obviously on the other projects' analyses. Especially, automated case handling (ACH) and its development and transformation/implementation have been specified in previous tables. The RQs of Stockholm are focused on one mentioned digital technology (RPA); see the Halmstad project for process automation technology and such examples (Table 4).

6 Synthesis and Conclusions: Clarifying the Subject Matter of Automated Case Handling Practices

6.1 Comparative Reflections

This study has investigated five research projects and their research questions on process automation. The main purpose of the chapter is not to make a comparative review of these research projects. The inquiry of RQs aimed mainly at clarifying practices and phenomena in such practices for process automation research. The use of five projects aimed at obtaining a *rich view of the landscape of practices and phenomena* related to process automation as described in Sect. 5. However, based on this inquiry, I make some comparative reflections on what has been clarified in the investigation of focused research phenomena. All projects have an aim to clarify the subject matter of process automation; how to *characterize* this kind of practice. The RQs can be seen as a *mixture of focusing* (through mentioning *certain phenomena*) and being relatively *open-minded* through fairly *general concepts*. The Linköping project aims at discovery and clarification of *meanings* and *consequences* of process automation. The Göteborg project talks about *effects* and the Örebro project talks about *influences*. The Halmstad project about *implications*. The Stockholm project addresses *changes* through process automation.

Two research projects are specific concerning what digital technologies to study for process automation; Halmstad on RPA and AI and Stockholm on RPA. The other projects do not in their RQs mention any specific technologies. Some of the projects apply public administration concepts in their RQs. Göteborg uses concepts of *discretion* and *law-proof decision-making*. Örebro uses concepts of *discretion, accountability,* and *legitimacy*. Halmstad uses the concept of *good bureaucracy*. In several of the projects, different aspects of governance are included. Göteborg states an interest in *value positions* and Stockholm mentions *goals*. Other projects address governance more implicitly.

All projects aim at *description, clarification,* and *explanation* of process automation. There are some projects that besides this take a more explicit *prescriptive* orientation by aiming at *guidelines* for different aspects of process automation (Linköping, Halmstad, and Stockholm).

6.2 Practice Ontology of Automated Case Handling

The analyses of the RQs in the research projects with clarification of practices and phenomena (Sect. 5) have produced an elaborate basis for a *concerted summary description of the subject matter of process automation*. Figure 4 is an ontological clarification of automated case handling that is based on the practice-theoretic view as described in Sect. 4.1.

There is in this "*ontological map*" a focus on an ACH practice, but this is made contextually that shows relationships to other practices. An automated case handling practice should not be interpreted as something fully automated without any human intervention or surveillance. There may be only parts (e.g., a restricted set of activities) of a case handling process that is automated. There may also be a differentiation between simple cases (which are handled automatically by digital artifacts) and complex cases (which need human treatment). This means that a case handling practice will be conducted by both human agents and digital agents. In such a practice, there will be automated activities, which are performed fully by a digital artifact (see Table 1).

From a practice perspective, there is a focus on the *performance* of case handling activities either they are conducted by digital agents or humans. The performances of case handling are based on what is here called *dispositions for actions*. Such dispositions comprise knowledge of diverse and related kinds: conceptual linguistic (what kind of practice phenomena and what to call them), performative (what actions and how to perform them), regulative (rules for actions and results), relational (what actors and their roles) and also normative foundations (ideals, values, and goals).

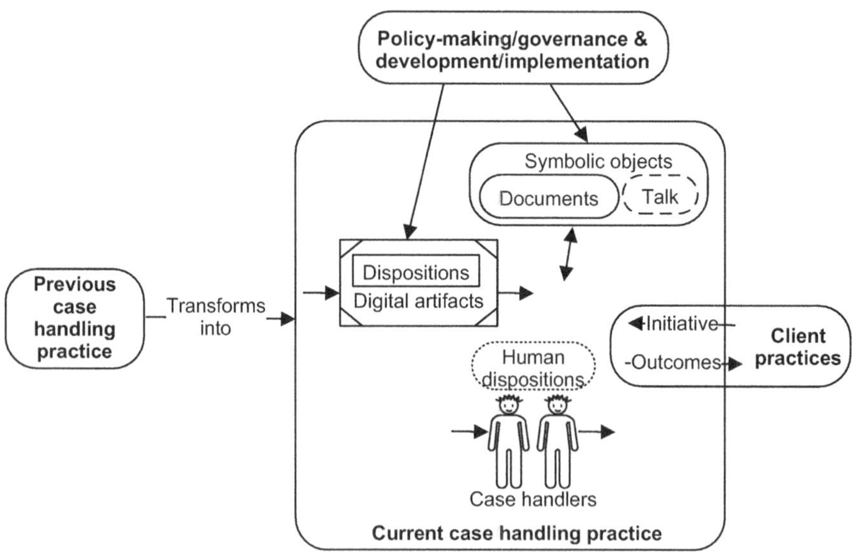

Fig. 4 Ontological clarification of an automated case handling practice in context

Dispositions cover thus several social phenomena mentioned in the ontological analysis in Sect. 5, such as competencies, role ascriptions, value positions, and goals.

The different agents in a case handling practice are *carriers* of such practical knowledge/dispositions, otherwise, without such *capabilities,* they could not perform those desired actions. The development and implementation of process automation mean that existing dispositions (within a "previous case handling practice") are refined, modified, and transformed into dispositions of a new/altered case handling practice. Digital dispositions must be inscribed into the software of the digital artifacts used for automated activities. Human dispositions originate from the humans' knowledge and experiences from previous case handling. It is important to acknowledge that dispositions are dispersed among different agents (digital and human) within such a practice. This means that there does need to be total equivalence between dispositions of such carriers.

There will also be a communicative influence from development and implementation concerning changes in procedures. This pertains especially to relational changes (roles); what human caseworkers do and what digital artifacts do. Such influences will be made communicatively through "symbolic objects," i.e., both documents and talk (oral communication). Examples are process models, produced during development, that describe and prescribe altered case handling processes. There is also a general influence from governance activities, such as policy-making and prescriptive development, on case handling (e.g., value positions, regulations, and guidelines). Such influences are also conducted communicatively.

In Fig. 4, it is also depicted the important relationships to client practices. Case handling is based on demands/initiatives for the handling of cases, and it should respond through decisions and other outcomes directed to such client practices. An ACH practice is, as a *transactional practice*, both (1) an *actional accomplishment* and *transformation* of a case (from an initiative to a decision) and, (2) an *interactional change of social relationships* between public authority and citizens (Goldkuhl & Röstlinger, 2006).

The different phenomena, ontologically clarified in Tables 1, 2, 3, 4, and 5, can be positioned within the overarching and synthesized description of automated case handling (Fig. 4). The ontological map shows the subject matter of process automation and it can be used for reasoning about how to conceive and address these phenomena.

6.3 Concluding Remarks on Chapter Contributions

The main contribution of this chapter is a clarification of the subject matter of process automation. This inquiry was made "indirectly" by the use of descriptions from other research projects and not based on specific empirical inquiries. The research questions of five research projects on process automation have been investigated. The contributions can be divided into (1) a *structure of* different *practices* related to process automation (especially Fig. 3), (2) *illustrations of* such *practices* through analyses of RQs in studied research projects (Sects. 5.1–5.5), (3) *ontological clarification* of different focused *phenomena* within such practices (Sects. 5.1–5.5, especially Tables 1, 2, 3, 4 and 5), (4) a *synthesized ontological map* of the subject matter of process automation based research project analyses (Sect. 6.2, especially Fig. 4), and (5) also a *foundational linguistic analysis* of "automation" and related concepts (Sect. 3; especially Fig. 2). What has thus been achieved is a *common conceptual* and *linguistic ground* that enables relating and comparing different research endeavors.

Besides these main contributions to process automation, I would also like to point out the *inquiry approach* taken for these analyses. I have conducted a *linguistic analysis* of primary concepts within the research territory of process automation. I have also conducted a *practice-ontological analysis* of research questions leading to the clarification of practices and phenomena within such practices. These conceptual instruments, illustrated in this chapter, should be helpful for scholars in planning their research endeavors and formulating pertinent research questions.

Acknowledgments I am most grateful to the researchers responsible for the inquired research projects: Ida Lindgren (Linköping), Agneta Ranerup (Göteborg), Karin Hedström (Örebro), Maria Åkesson and Michel Thomsen (Halmstad), and Gustaf Juell-Skielse (Stockholm). I have received useful knowledge through interviews and written descriptions.

References

Abbott, A. (1992). From causes to events. Notes on narrative positivism. *Sociological Methods & Research, 20*(4), 428–455.

Åkesson, M., & Thomsen, M. (2020). RPA in service of bureaucracy – When bots are colleagues in everyday public administration. In *The 17th Scandinavian Workshop on E-Government* (SWEG-2020), Göteborg.

Alvesson, M., & Kärreman, D. (2000). Taking the linguistic turn in organizational research. Challenges, responses, consequences. *Journal of Applied Behavioral Science, 36*(2), 136–158.

Andersson, A., Hedström, K., & Wihlborg, E. (2018). Automated decision-making and legitimacy in public administration. In *The 15th Scandinavian Workshop on E-Government* (SWEG-2018), Copenhagen.

Dewey, J. (1938). *Logic: The theory of inquiry*. Henry Holt.

Eastwood, J. (1994). *Oxford guide to English grammar*. Oxford University Press.

Ehn, P., & Kyng, M. (1984). A tool perspective on the design of interactive computer support for skilled workers. In M. Sääksjärvi (Ed.), *Report of the Seventh Scandinavian Research Seminar on Systemeering*. Helsinki.

Emirbayer, M. (1997). Manifesto for a relational sociology. *American Journal of Sociology, 103*(2), 281–317.

Goldkuhl, G. (2002). Anchoring scientific abstractions – Ontological and linguistic determination following socio-instrumental pragmatism. In *Proceedings of European Conference on Research Methods in Business*. Reading.

Goldkuhl, G. (2005). Socio-instrumental pragmatism: A theoretical synthesis for pragmatic conceptualisation in information systems. In *Proceedings ALOIS-2005*. University of Limerick.

Goldkuhl, G. (2019). The generation of qualitative data in information systems research: The diversity of empirical research methods. *Communications of AIS, 44*, Article 28.

Goldkuhl, G., & Röstlinger, A. (2006). Context in focus: Transaction and infrastructure in workpractices. In *Proceedings ALOIS-2006*. Borås.

Güner, E., Han, S., & Juell-Skielse, G. (2020). Robotic process automation as routine capability: A literature review. In *Proceedings ECIS-2020*.

Houy, C., Hamberg, M., & Fettke, P. (2019). Robotic process automation in public administrations. In M. Räckers et al. (Eds.), *Digitalisierung von Staat und Verwaltung, Lecture Notes in Informatics (LNI)*. Gesellschaft für Informatik.

Lacity, M., & Willcocks, L. (2016). Robotic process automation at Telefónica O2. *MIS Quarterly Executive, 15*(1), 21–35.

Lascoumes, P., & Le Gales, P. (2007). Introduction: Understanding public policy through its instruments—From the nature of instruments to the sociology of public policy instrumentation. *Governance, 20*(1), 1–21.

Lindgren, I. (2020). Exploring the use of robotic process automation in local government. In *EGOV-CeDEM-ePart 2020*.

Lindgren, I., Toll, D., & Melin, U. (2021). Automation as a driver of digital transformation in local government. In *Proceedings Digital government 2021*. ACM.

Mahoney. (2005). The histories of computing(s). *Interdisciplinary Science Reviews, 30*(2), 119–135.

Nicolini, D. (2012). *Practice theory, work, & organization*. Oxford University Press.

Olson, M., & Lucas, H. (1982). The impact of office automation on the organization: Some implications for research and practice. *Communications of ACM, 25*(11), 838–847.

Ranerup, A., & Henriksen, H. Z. (2019). Value positions viewed through the lens of automated decision-making: The case of social services. *Government Information Quarterly, 36*(4)., Article 101377.

Ranerup, A., & Henriksen, H. Z. (2020). Digital discretion: Unpacking human and technological agency in automated decision making in Sweden's social services. *Social Science Computer Review*, 1–17.

Reckwitz, A. (2002). Toward a theory of social practices. A development in culturalist theorizing. *European Journal of Social Theory, 5*(2), 243–263.

Ryle, G. (1949). *The concept of mind*. Hutchinson.

Schatzki, T. R. (2001). Introduction: Practice theory. In T. R. Schatzki, K. Knorr Cetina, & E. von Savigny (Eds.), *The practice turn in contemporary theory*. Routledge.

Wihlborg, E., Larsson, H., & Hedström, K. (2016). "The Computer Says No!" – A case study on automated decision-making in public authorities. In *HICSS-2016*. IEEE.

Understanding Automated Decision-Making in the Public Sector: A Classification of Automated, Administrative Decision-Making

Ulrik B. U. Roehl

1 Introduction

Service automation in the public sector is applied to a range of different activities such as policy development, public service delivery, internal management and administrative decision-making. Each activity shares common traits and is characterised by particularities of use and technology. This chapter focusses on semi- and fully automated administrative decision-making (AADM[1]) utilised by public administrative bodies. AADM is here defined as administrative decision-making being partly or fully based on automated outputs generated by algorithmic systems that incorporate relevant regulation of a given policy area.

Administrative decision-making is the unilateral determination by public administrative bodies regarding what is or what shall be lawful in specific cases. This determination is based on both attributes and relevant legislation and made in relation to an individual citizen or firm. Globally, millions of such decisions are taken every day in policy areas such as administration of traffic offences, allocation of grocery market stalls, taxation, social security benefits and child abuse prevention. Empirically, it is widely assumed that the use of AADM has increased in the public sector due to technological advances and will continue to do so.

Automated decision-making including AADM has been discussed by multiple authors in terms of, for example, efficiency (e.g. Vogl et al., 2020), quality (e.g. Kuziemski & Misuraca, 2020), accountability (e.g. Smith et al., 2010),

[1] "ADM" is a common abbreviation for automated decision-making; "AADM" is used in this chapter to emphasise the focus on automated administrative decision-making as a particular type of automated decision-making.

U. B. U. Roehl (✉)
Center for Organization, Management and Administration, Department of Politics and Society, Aalborg University, Aalborg, Denmark
e-mail: uhr@dps.aau.dk

G. Juell-Skielse et al. (eds.), *Service Automation in the Public Sector*, Progress in IS,
https://doi.org/10.1007/978-3-030-92644-1_3

transparency and judicial review (e.g. Cobbe, 2019), professional discretion (e.g. Buffat, 2015), norms of civil servants (e.g. Wihlborg et al., 2016) and role of system designers (e.g. Zouridis et al., 2020). Most of those studies refer to intuitive but simplified categories of automated decision-making based on a 3-fold differentiation between no automation, semi-automated decision-support and fully automated decision-making.

While the contributions offer important understanding regarding consequences of automation, the understanding of AADM tends to be simple, and studies are often derived from formal understandings of technology instead of its actual use in organisational settings (Peeters, 2020). This jeopardises comparisons of AADM usage due to a lack of common and precise definitional base. These contributions do not cast much light on "areas in between" the three simplified categories although those areas have been pointed out as important for future research (Busch & Henriksen, 2018; Lange et al., 2019). This tendency further risks leading to methodological inconsistencies as constructivist approaches to technological use are often accompanied by more deterministic approaches to technology itself.

Asking how to best conceptualise AADM usage in order to understand its wider consequences for the public sector and society, a fine-grained classification of six ideal types of use of AADM is suggested. Drawing on key references within the academic disciplines of Public Administration, Decision-support Systems and Science & Technology Studies, each type describes a configuration of decision authority between civil servants and algorithmic systems.

The chapter is based on an understanding of empirical applications of AADM as examples of wider algorithmic systems being grouped together by combinations of multiple systems, government databases, citizen portals and network components (Nevo et al., 2009; Stoudt-Hansen et al., 2020). Such systems include, but are not limited to, techniques such as robotic process automation, rule-based (expert) models, regression, big data, predictive analytics, machine learning and neural networks which are accessed by civil servants through the operation of smart phones, tablets, websites, office applications and case management systems (see Busch & Henriksen, 2018, for use of some of the latter applications).

Primarily due to technological progress, some authors have argued that the role of individual civil servants in administrative decision-making is "doomed" in the long run (Zouridis et al., 2020). Nonetheless, this chapter is based on a belief that different configurations of shared decision authority between civil servant and algorithmic systems will be with us for a long time due to technological, organisational, political and ethical issues.

The chapter builds on broad sociotechnical understandings of humans and technology and stresses how technology frames human possibilities for action but does not determine the action (Lips, 2020; Plesner & Husted, 2020). Each configuration of decision authority thus illustrates how technology usage does not exist independent of its users and cultural context around it and thereby further explore this volume's underlying combination of social and technical perspectives. Rather than implying an inevitable development towards fully automated administrative decision-making, the classification allows for understandings of multiple co-existing empirical types of AADM usage and stresses the importance of civil servants' sensemaking and interpretations as well as cultural context.

Contributing to emerging literature on automated decision-making and public service automation, the chapter provides for new sensitivities to empirical applications beyond the basic differentiations of no automation, semi-automated decision-support and fully automated decision-making. The suggested classification has practical relevance by helping to identify and understand similarities and differences in the use of AADM across organisational settings and policy areas thereby supporting informed choices of appropriate system design and test as well as choices of appropriate professional and management practices in relation to AADM usage.

The chapter proceeds as follows. Firstly, the concept of administrative decision-making is explored in detail including current literature on the automation of such decisions. This is followed by a discussion of the methodological basis of the suggested classification and a review of relevant, existing definitions, classifications and typologies. The main part of the chapter is the development and discussion of the proposed classification. Before concluding, the usefulness of the classification for practice and future research is discussed.

2 Automated Administrative Decision-Making

Administrative decision-making is the everyday activity of public sector bureaucracies and involves a large number of civil servants and case workers within the executive branch at all levels of government worldwide.[2] It is here understood as the unilateral determination by a public administrative body—through a formal decision, administrative act or adjudication[3]—of what is or what shall be lawful in specific cases based on its attributes and relevant statutory regulation (including possible underlying legislative guidance) and in relation to an individual citizen, a firm or a group of those (Mashaw, 2007; Stelkens, 2020).

A distinctive characteristic of administrative decision-making compared to other types of service automation in the public sector is the surrounding legal framework. Specifically, while the decisions themselves are based on specific statutory regulation such as a clean air act (and subjacent government orders, etc.), administrative decision-making takes place within a procedural, legal framework in terms of administrative legislation and standards of good administration which emphasises elements such as due process, contradictory procedures, accountability, obligation of

[2] "Civil servant" is used as a term for case workers, case managers, adjudicators and other officials who are responsible for administrative decisions. In addition and for sake of ease, the singular "civil servant" is used although often it is empirically more correct to speak of civil servants in plural.

[3] Although frameworks of administrative law vary across legal traditions, the concept of administrative decisions is generic and known under headings such as "acte administratif individual" (Francophone tradition); "Verwaltungsakte" (German tradition); and "förvaltningsbeslut"/"-afgørelse"/"-vedtak" (Scandinavian tradition).

reason giving, equality of treatment and the principle of proportionality (Bell, 2006; Widlak et al., 2021). It follows that administrative decision-making is often based on specific and rather strict procedural requirements with the administrative decision being "the end point" (Eberle, 1984) of the decision-making process.

What particularly signifies administrative decisions is their individual, "definite" legal character (with the possibility for later modification via formal complaint, review and appeal procedures). Following case-handling steps, the decision settles a case by determining what is or what shall be lawful for the involved parties based on relevant statutory regulation.

Administrative decision-making covers an extensive spectrum of activities: some beneficial to the individual (e.g. decisions to grant unemployment benefit or children benefits) and some restrictive (e.g. denial of permission to build a house or denial of parole). While some administrative decisions are not particularly important, many have serious consequences concerning, for example, eligibility for social security benefits or limits on firms' environmental emissions.

The complexity of administrative decision-making differs from simple decisions on speeding fines primarily based on a single attribute (speed of the driving vehicle) and one legal aspect (speeding limits) to more complex decisions such as the assessment of permitted emissions of hazardous pollutants from an industrial polluter based on clean air legislation. These decisions are based on multiple attributes of the case as well as several legal aspects.

Herbert A. Simon's (1960) classic categories of decisions illustrate this. He argues that decisions range from highly structured to semi-structured and highly unstructured decisions. Structured decisions refer to routine and repetitive problems for which solutions are well known, while unstructured decisions are unclear and characterised by no obvious solutions. Semi-structured decisions occur when some (but not all) elements are structured (Averweg, 2010).

In his work, Simon (1960) establishes three generic phases of decision-making which can be approximately applied to administrative decision-making. In the first phase (intelligence), data relevant to the decision is compiled and assessed. In the second phase (design), possible courses of action are developed, and in the third phase (choice), a particular course of action is chosen. Simon (1960) points out the phases are not necessarily linear just as they might be more or less formalised. Most administrative decision-making will build on rather formalised phases: An initial assessment of the attributes of the case in question; secondly, a series of procedural steps to develop possible decisions; and lastly the application of statutory regulation to the individual case in order to reach the actual administrative decision.

Combining Simon's concepts, it is possible to imagine administrative decisions which differentiate in terms of complexity across the three phases. In general, most administrative decisions are structured or semi-structured as their statutory basis to some extent stipulates the attributes to be taken into account, the necessary procedural steps to be taken and the range of possible decisions to be considered. Related to the complexity of administrative decisions is the scope of administrative discretion of the civil servant (Rosenbloom et al., 2010). All other things being equal, the more complex the decision is, the more likely the need for administrative discretion.

Empirically, it is widely assumed that the automated administrative decision-making usage has increased in the public sector due to technological advances. A report released by the Swedish National Audit Office in 2020 counted 112 automated decision processes within 13 Swedish central government agencies. In total, an estimated 137 million annual automated administrative decisions were made, of which 121 million were fully automated (Riksrevisionen, 2020). Generally though, few quantitative assessments of the extent of AADM usage exist.

Researchers have instead used case studies to explore the AADM usage. Appendix details 10 exemplary studies of AADM usage which show variation across policy areas and national settings. These studies include both semi- and fully automated administrative decision-making within areas such as minor traffic offences, correctional services, child protection, driving license permits and social security benefits in Australia, Europe, and the USA.

Despite their increasing number and empirical variation, such studies seldomly carry information on technological usage beyond the above-mentioned, simplified 3-fold differentiation. These studies seldomly describe the distribution of decision authority between civil servants and technology or the mutual influence between the latter two. Although most authors are careful to state the scope of theirs claims, a reader can be led to the conclusion that fully automated decision-making based on big data and artificial intelligence is the new normal, serving as an inevitable reference point for AADM usage. Read carefully, however, the studies reveal different patterns of AADM usage as well as algorithmic systems including different combinations of specific techniques.

3 Methodological Basis

This chapter is conceptual and proposes new links across disciplines and associations that refine how to consider technology usage in administrative decision-making among administrative bodies in the public sector. AADM usage is what Jaakkola (2020) calls the focal phenomenon of the suggested classification. Instead of an empirical-based study or test of the new links, this chapter focusses on developing logical and comprehensive arguments (Gilson & Goldberg, 2015) and organising existing research into common distinct types (Jaakkola, 2020) via a classification. Classifications are the result of a process of "...ordering entities into groups or classes on the basis of similarity" (Bailey, 1994). They serve as a tool for the advancement of research including the development of theories (Nickerson et al., 2013, building upon Iivari, 2007).

The classification developed here was inspired by the abductive method suggested by Nickerson et al. (2013). Based on knowledge of relevant literature as well as empirical instances of AADM usage, the process started with the identification of the configuration of decision authority between civil servants and algorithmic systems as the key differentiator of AADM usage which served as the "meta-characteristic" of the classification (Nickerson et al., 2013).

A structured literature search was then performed to identify possible existing classifications, typologies and taxonomies of automated decision-making, algorithmic decision-making, data-driven decision-making, decision-support systems and similar concepts (search terms consisted of variations of "classification" and "automated decision-making", respectively). Mirroring a deductive process—termed a conceptual-to-empirical approach by Nickerson et al. (2013)—this resulted in the initial, preliminary description of six ideal types. The search was performed through "Scopus" and "Web of Science" across English language sources in the categories of computer science, business, management & accounting and social sciences. No existing classification, typology or taxonomy of AADM or of AADM usage were identified. Instead, several generic classifications as well as classifications within particular contexts were identified and used as inspiration for the suggested classification.

Existing empirical and theoretical studies were scanned for explicit (e.g. Bovens & Zouridis, 2002) or implicit (e.g. Sun & Medaglia, 2019) descriptions of different types of AADM or different types of AADM usage. Through this inductive, empirical-to-conceptual approach (Nickerson et al., 2013), the six ideal types of the classification were further elaborated just as the understanding of the configuration of decision authority as the key differentiator was refined.

Clear principles on how to assess the validity of conceptual work within the social sciences are scarce. Even so, the proposed classification can be assessed based on what researchers have pointed to as qualities and advantages of conceptual work. Condensing the suggestions of Bailey (1994) and Nickerson et al. (2013) into four such criteria, the validity and usefulness of the classification will be discussed in section 6 of the chapter.

A final methodological note: the classification consists of ideal types and build on the tradition of Max Weber (1904/2012). Ideal types are constructs and not empirical entities (Bailey, 1994). While the six ideal types are "empirically plausible" and constructed to be the *clearest* illustrations of empirical instances of AADM usage, they are by principle unlikely to match any specific, empirical example of AADM usage in detail.

4 Insights from Existing Definitions, Classifications and Typologies

In 1977, Steven Alter suggested seven generic categories of so-called decision-support systems which have been particularly influential in the academic field of Decision-support Systems (Power, 2007). Alter's category of systems, based on suggestions models, is a good starting point for understanding AADM and highlights the centrality of decisions. Systems as discussed by Alter "perform mechanical work leading to a specific suggested decision for a fairly structured task" (Alter,

1977, p. 42) and are based on "specialized problem-solving expertise" within a particular domain (Power, 2004, p. 162).

Only a handful authors have attempted to define or describe AADM (including similar concepts such as administrative algorithmic decisions) more precisely. Some of those definitions have focused on technology and particularly stress the use of machine learning techniques (Cobbe, 2019; Oswald, 2018). Conversely, functional definitions stress how AADM includes the automated compilation, processing and application of information as the basis of administrative decisions. Of the latter, some suggestions are broad and hardly include particular characteristics of administrative decision-making (Schuilenburg & Peeters, 2021), while others stress the importance of administrative decisions being based on statutory regulation and agency guidance and procedures (Hogan-Doran, 2017; Widlak et al., 2021; Wihlborg et al., 2016).

A few authors point to roles of civil servants and algorithmic systems exemplifying different configurations of responsibility (Widlak et al., 2021; Wihlborg et al., 2016) which mirrors Alter's (1977) more general emphasis on the "degree of action implication of system outputs (i.e., the degree to which the system's output could directly determine the decision)" as a key variable.

There seems to be growing consensus in academia that instances of AADM can be placed on a continuum of automation from automation which provides different types of guidance to the civil servant to fully automated decision-making which leaves no role for the civil servant. The two basic end points of this continuum are somehow mirrored by the popular phrases of the human operator being either "in" or "out" of the decision-loop[4] which points to the basic concepts of semi and fully automated decisions described early in the chapter.

To develop the proposed classification, the following subsections will discuss five existing classifications and typologies of automation in public administrative settings and beyond as presented in Table 1. The table maps the classifications according to the simplified 3-fold differentiation as shown in the left column. Taken together, it is possible to shed valuable light on the aforementioned automation continuum. It is important to note that the classifications are not—despite the seemingly orderly appearance of the table—fully comparable due to differences in definitions, focal phenomena, ontology and epistemology.

[4]Originating in relation to autonomous weapon systems, industrial production, etc., and occasionally mentioned as a theoretical possibility in discussions of automated decision-making in the public sector, it is also possible for the human operator to be 'on' the loop. This implies the operator is supervising the fully automated decision-making with the ability to stop it within a given timeframe (Hauptman, 2013). Empirical instances of the "on"-type in relation to administrative decision-making seem to be very few or non-existent.

Table 1 Overview of existing classifications and typologies (explicit or implicit) of automation. Types are not fully comparable across sources. "AI-based" denotes automated decision-making particularly based on artificial intelligence techniques

Simplified type	Bovens & Zouridis (2002)	ORAD Committee (2021)	Parasuraman et al. (2000)	Bader & Kaiser (2019)	Lange et al. (2019)
Approach	*Technology-centred*			*Sociotechnical*	
Focal phenomenon	*Administrative decision-making in large executive agencies*	*Driving automation of motor vehicles*	*Human interaction with automation (generic)*	*AI-based decision-making in call-centre*	*AI-based high-frequency trading.*
No automation	Street-level bureaucracy	No driving automation	Not covered	Not covered	Not covered
Semi automated decision-support ("in-the-loop")	Screen-level bureaucracy	Driver assistance	Information acquisition	High user involvement; attachment to decision Low user involvement; detachment from decision	Algorithms as objects
		Partial driving automation	Information analysis		Algorithms as quasi-objects
			Action and decision selection		Algorithms as quasi-subjects
Fully automated decision-making ("out-of-the-loop")	System-level bureaucracy	Conditional driving automation	Action implementation	Not covered	Algorithms as subjects
		High driving automation			
		Full driving automation			

4.1 Technology-Centred Understandings

In 2002, Bovens and Zouridis introduced a differentiation between "street-level", "screen-level" and "system-level" bureaucracies in what the authors called large, public "decision-making factories". This mirrors the basic 3-fold differentiation between no automation, semi-automated decision-support and fully automated decision-making and is the explicit or implicit departing point for many studies regarding automated decision-making in the public sector including specific studies of AADM (Bannister & Connolly, 2020, for example, use the terms "passive" and "active" algorithms mirroring the latter two categories).

Looking beyond analyses of decision-making in the public sector, work on self-driving vehicles can serve as further inspiration. The global engineering association, SAE International, has developed a standard of such vehicles (On-Road Automated Driving (ORAD) Committee, 2021) and maps six types of automation. What is interesting here is the fine-grained nature of the standard. Besides an initial type of no automation, the classification's "Driver assistance", "Partial driving automation" and "Conditional driving automation" are detailed examples of the driver gradually entrusting more responsibility to the vehicle (On-Road Automated Driving (ORAD) Committee, 2021).

Thomas B. Sheridan has worked with human–automation interaction for several decades and originally suggested a detailed classification of automation based on 10 "degrees" of automation independent of any particular type of technology (Sheridan, 1992). The classification has since been simplified to five levels (none, low, medium, high and full automation) and combined with a functional dimension of automation.[5] In functional terms, Sheridan and his associates differentiate between (i) information acquisition; (ii) information analysis; (iii) action and decision selection and (iv) action implementation (Parasuraman et al., 2000).

"Information acquisition" describes the automated compilation and registration of data that supplement the human operator's information search and selection of a course of action. "Information analysis" describes the automated configuration and presentation of data for the human operator and supports the human interpretation of data. An example of this could be that a key parameter is above a certain threshold (e.g. the threshold being a speed limit for cars or a risk indicator for child abuse). A defining characteristic is that "information analysis" does not include any recommended courses of action.

"Action and decision selection" describes the automated selection among decision alternatives. This could, for example, be an automated system designed to perform a specific decision choice if particular conditions exist. Based on Sheridan's original classification, this selection could take several forms from automated narrowing down of multiple decision choices to a few options to automated execution of a decision after a certain timeframe if the human operator has not chosen otherwise. "Action implementation" describes situations where decisions are taken in an automated manner mirroring "fully automated" as used in this chapter (Parasuraman et al., 2000).

4.2 Sociotechnical Understandings

The classifications discussed so far have primarily focussed on technology itself whether described as three, five or 10 "degrees" of automation. It is, therefore, meaningful to briefly consider previous work which strengthens the understanding of automated administrative decision-making usage based on broad sociotechnical understandings.

The actual use of technology is seldomly identical to the intended use. Contextual factors such as situational constellations, practical contexts, organisational structures (Lange et al., 2019), surrounding legal framework and individual traits of possible human operators affect technology usage including AADM. Peeters (2020) points out how "...bounded rationality, satisficing behaviour, automation bias and frontline coping mechanisms play a crucial role in the way humans make use of the oversight and override options built into algorithms". Fully understanding the use of technology (including AADM) thus requires a focus on the sensemaking and interpretations

[5] Parasuraman et al. (2000) suggest a two-dimensional model of function <u>and</u> level of automation. For sake of clarity and adaptability to administrative decision-making this has been combined to one dimension in Table 1

surrounding the technology (Liu & Graham, 2021) and cultural context (Plesner & Husted, 2020).

Rather than fixed configurations of responsibility between civil servants and algorithmic systems according to a priori (system) design, inspiration from sociotechnical approaches gives reason to expect a mixed picture of configurations as much based on the technology as on users and contextual factors surrounding usage. It furthermore gives reason to reconsider the meaning of "use" as this concept should be understood as multiple *practices* of civil servants and others in relation to AADM rather than solely tangible commands, instructions and messages between civil servants and algorithmic systems. While these practices evolve around technology, they are not limited to it (Bailey & Barley, 2020).

In their study of the use of an automated decision-tool based on artificial intelligence techniques in a commercial call centre supporting sales activities, Bader and Kaiser (2019) emphasise how the tool gave rise to differing and dynamic forms of joint problem-solving between human operators and technology. As shown in Table 1, the authors describe a continuum between attachment to decisions based on different elements of user involvement and detachment based on "spatial and temporal separation, rational distancing, and cognitive displacement" in relation to the operations of the technology. It is important to note that use of the same technology in the same empirical setting can give rise to instances of both low and high involvement due to human operators and organisational conditions (Bader & Kaiser, 2019).

Lange et al. (2019) in their examination of high-frequency trading in financial markets provide an interesting example of classification based on different perceptions of "subject-object relations" between individual traders and what the authors call "trading algorithms". Via an ethnographic approach, they trace four types of practices ranging from perceiving algorithms as objects (i.e. the trader controls the algorithms) to the trader seeing her/himself as a tool for partly independent algorithms. What is interesting here is how the traders perceive themselves, and—most likely—act in relation to the technology. Rather than being able to clearly detach the role of human operators (be it trader or civil servant) from the technology, Lange et al. (2019, p. 611) suggest to "...shift the attention away from the extremes ("warm intentions" or "cold codes") to the areas in between where both extremes merge, sometimes becoming seemingly indistinguishable".

Based on these insights, a classification of AADM usage must allow for a thorough understanding of medium forms of automation—"the areas in between"—and perhaps even accept that practices of each individual civil servant vis-à-vis an algorithmic system can potentially be characterised by a dynamic and unique configuration of decision authority (Veale et al., 2018). In other words, the distribution of decision authority across civil servant and algorithmic system might—even when speaking of the same algorithmic system—change depending on the individual and time.

5 A Classification of Use of Automated, Administrative Decision-Making

Building on the previous sections, this section introduces the suggested classification of AADM usage. The classification defines six ideal types of AADM usage ranging from Minimal automation (Type A) to Automated decisions (Type E) and Autonomous Decisions (Type F) and describes an increasing reliance on automated output in the decision-making process. It is illustrated in Fig. 1 and detailed in Table 2.

It is important to stress the methodological openness of the suggested classification: the intention is to provide a tool to describe and analyse AADM usage. The classification is not a normative statement on desired levels of automation in the public sector, nor is it based on any assumption that all administrative decision-making will eventually evolve into autonomous decisions based on advanced artificial intelligence techniques.[6]

The classification is functional as it describes technology usage rather than technology itself. It follows that advanced techniques such as predictive analytics, machine learning or data mining could well be enshrined in algorithmic systems of the semi-automated types of B, C and D as they are in the fully automated types of E and F.

The classification is a descriptive tool intended to be applied to automated, administrative decisions of *differing complexity*. Departing from the ideas of Simon (1960) outlined earlier, one can, however, predict that Automated decisions (Type E) and Autonomous decisions (Type F) are more likely to involve highly structured, administrative decisions. Moreover, Simon's two initial decision-making phases of intelligence and design are—other things being equal—probably easier to automate than the third, choice, phase. This implies that in instances where semi-structured or unstructured administrative decisions are actually subject to automation, technology most likely has the role of compiling, registering and presenting data for the civil servant (mirroring Type B) and possibly suggesting appropriate procedural steps (mirroring Type C).

Before scrutinising the six types in greater detail, the conceptual background must briefly be considered. The classification defines ideal types each representing a configuration of *decision authority* between civil servants and algorithmic systems. Authority can be understood in several ways (Bourgoin et al., 2020) and is here related to the idea of authority as acceptance: authority covers the explicit or implicit "... right to decide on specified matters to a member or group of members of the organization" (Aghion & Tirole, 1997). Applied to administrative decision-making,

---

[6] A short caveat is appropriate in relation to the illustration of the classification (Figure 1): the illustration is downward sloping towards Autonomous decisions (Type F) thereby risking indicating a negative understanding of this type of automated decision-making (i.e. towards a "digital nightmare"). Bearing the descriptive nature of the classification in mind, this is not the intention, but the sloping character has been chosen—as a matter of the lesser of two evils—to avoid the risk of indicating a positive understanding of a "digital nirvana" through an upward slope.

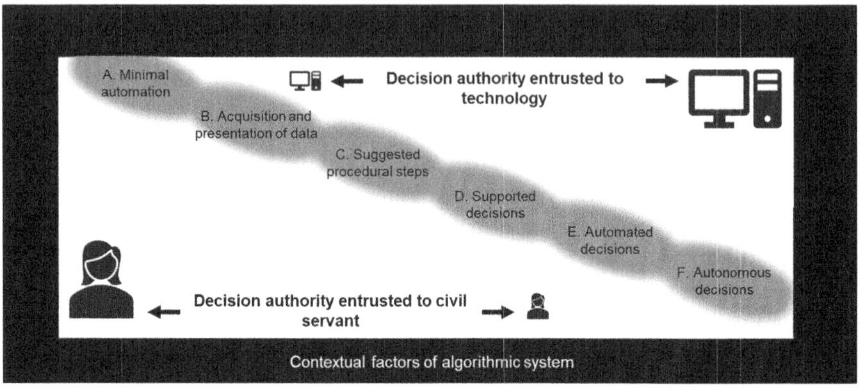

Fig. 1 Illustration of classification of use of automated administrative decision-making (AADM)

different configurations of decision authority can be classified by the degree to which AADM usage entails the determination of the administrative decision by automated output of the algorithmic system. This mirrors the idea of different "grades" of shared responsibility between human operators and technology as discussed in the previous section.

The classification furthermore focusses on *organisational practices* in relation to AADM within a given administrative body and in a particular policy area (i.e. traffic offences or air pollution control). Instead of describing the intended or specified use of a given IT-system, the classification allows for mapping of the decision-making practices surrounding algorithmic systems. If civil servants exhibit an over-reliance on automated suggestions for decisions and are not exercising individual assessment as understood in the concept of automation bias (Cummings, 2006), organisational practices can be classified as Automated decisions (Type E) rather than Supported decision (Type D).

It is appropriate to understand AADM usage as unfolding within what some authors have referred to as *algorithmic systems* (e.g. Kellogg et al., 2020). While Seaver (2019) perceives such systems as "arrangements of people and code", they are seen as more or less complex combinations of technologies (and not people) here. Instead of perceiving AADM usage as based on one particular IT-system and one particular technology (e.g. machine learning), civil servants operate several interfaces connected to multiple systems, databases, citizen portals and network components constituting "bureaucratic information architectures" (Peeters & Widlak, 2018) as part of the automated decision-making process. While the interfaces operated by civil servants might be stable over periods of time, the connectedness of algorithmic systems mean they are open-ended in principle and changing as tiny parts are constantly tweaked, tuned and swapped (Seaver, 2019).

5.1 Six Ideal Types of Use of Automated, Administrative Decision-Making

Table 2 further details the six types of the suggested classification. The broad 3-fold differentiation found in much of the literature is roughly mirrored in the table: no or very limited automation corresponds to Minimal automation (Type A); semi-automated decision-support corresponds to the three types B, C and D, while fully automated decision-making corresponds to Automated decisions (Type E) and Autonomous decisions (Type F).

Building on Table 2, it is possible to elaborate on the characteristics of the six types.

Including Minimal automation (Type A) in a classification of *automated* administrative decision-making might initially appear contradictory; however, in public administrative contexts where the classification is empirically relevant, it is unlikely to encounter administrative decision-making not supported by simple technologies such as word processing at some point. Although limited in depth and scope, simple technologies in principle also support and shape collective practices. Additionally—and this is a characteristic shared with the semi-automated types B, C and D—it is highly likely that these practices are mutually supported and shaped by written check-lists, decision-rules and regulation (referred to as "written standards" in Table 2) reminding us that technology is not the sole factor formalising behaviour and limiting discretion within administrative bodies (Schartum, 2018).

Although inspired by the work of Sheridan, it should be noted that the classification strictly differentiates between Suggested procedural steps (Type C) and Supported decisions (Type D). While the former implies the civil servant taking guidance on the appropriate processual step(s) from the technology, the latter implies the civil servant is provided with recommendations of one or more possible decisions from a group of possible decisions.

Drawing on the empirical examples in Appendix, Fahnøe (2015) discusses the use of the DUBU system in the area of child protection in Denmark. Use of this system entails that civil servants are presented with selected data and led through procedural steps in order to manually assess the needs of protected children and decide on relevant interventions (the latter representing an administrative decision). The steps are meant to ensure compliance with statutory and budgetary requirements as well as professional standards of social work. While civil servants are presented with procedural requirements and options, the system does not suggest either a range of possible decisions or specific decisions (Fahnøe, 2015) and thus approximately mirrors Suggested procedural steps (Type C).

In contrast, Engstrom and Ho (2020) discuss the use of the QDD and Insight systems for the administration of disability benefits in the USA. Among other features, the systems compile, register and present relevant data of each case and automatically assess whether the case is what the authors term an "easy grant" to be approved without further assessment. In the event of such grants, civil servants are presented with the suggestion and are then meant to review and possibly approve the

Table 2 Ideal types of use of automated, administrative decision-making (AADM)

Simplified type	Ideal type	Description
No automation	A. Minimal automation	Civil servant has primary decision authority within the wider algorithmic system. Nearly all aspects of administrative decision-making are entrusted to civil servant and are solely supported by simple technologies such as word processing. Decision-making may be supported by written standards, etc.
Semi-automated	B. Acquisition and presentation of data	Civil servant and technology share decision authority within wider algorithmic system. Technology automatically compiles, registers and presents some or all data relevant to the case supplementing information acquired by civil servant. Remaining aspects are entrusted to civil servant. Decision-making may be further supported by written standards, etc.
	C. Suggested procedural steps	Civil servant and technology share decision authority within wider algorithmic system. Technology automatically compiles, registers and presents some or all data relevant to the case and suggests appropriate further procedural step(s). Remaining aspects are entrusted to civil servant. Decision-making may be further supported by written standards, etc.
	D. Supported decisions	Civil servant and technology share decision authority within wider algorithmic system. Technology automatically compiles, registers and presents some or all data relevant to the case and suggests a narrow range of decisions or a specific decision. Remaining aspects are entrusted to civil servant. Decision-making may be further supported by written standards, etc.
Fully automated	E. Automated decisions	Technology has primary decision authority within the wider algorithmic system. All aspects are entrusted to technology and performed automatically within static, explicit input–output relations and without support of civil servant.
	F. Autonomous decisions	Technology has primary decision authority within wider algorithmic system. All aspects of administrative decision-making are entrusted to technology and performed automatically within dynamic, implicit input–output relations (based on unsupervised learning techniques) and without support of civil servant.

decision (Engstrom & Ho, 2020) thus approximately mirroring Supported decisions (Type D).

A key differentiator of type C and D is based on the importance given to the final decision of what is or what shall be lawful in procedural, legal frameworks. Many duties of the administrative body as well as rights of the citizen or firm ultimately come into being in relation to the actual decision (e.g. obligation of reason giving) thereby clearly demarcating Suggested procedural steps (Type C) from Supported decisions (Type D).

In a similar vein, there is a fundamental difference between Supported decisions (Type D) and the two fully automated decision-making types of Automated decisions (Type E) and Autonomous decisions (Type F). The latter two describe organisational practices where the technology is relied upon to make and implement administrative decisions without any prior assessment by civil servants. As we know from other empirical settings, such practices might evolve due to routinisation and automation bias even though the technology itself needs a command from the human operator to finalise the decision-making process (Cummings, 2006, traces the history of several, high profile examples of over-reliance and automation bias). Civil servants might thus have the ability to review and override the decision at a later stage; however, the common, defining characteristic of those two types is the absence of continual, prior, human assessment meaning that the primary decision authority is entrusted with technology.

The defining difference between Automated decisions (Type E) and Autonomous decisions (Type F) is the nature of the input–output relations in the underlying decision models. Taken to the extreme, the difference has received considerable interest across disciplines, as Autonomous decisions potentially reflect fears of runaway algorithms based on advanced artificial intelligence techniques such as unsupervised learning (often contrasted to "old-fashioned" expert systems based on explicit if-then rules).

The understanding proposed here is a bit different. It is beyond doubt that machine learning and other artificial intelligence techniques which learn on the basis of patterns in data, necessitate a thorough discussion in terms of rule-of-law (Zalnieriute et al., 2019). Nonetheless it is also necessary to differentiate the degree of "intelligence" of the underlying decision models. It is entirely feasible to imagine semi-advanced decision models being inferred from historic patterns in the data by machine learning techniques but subsequently assessed and made explicit by humans before being put into operation.[7] This would effectively lead to the sharing of many of the same characteristics as seen in advanced expert systems. It is also feasible to imagine decision models based on historic patterns so advanced that they cannot be fully assessed by humans, just as decision models might continually to develop based on emerging patterns while in operation. The defining characteristic is thus the difference between static, explicit input–output relations represented in Automated decisions (Type E) and dynamic, implicit input–output relations (often termed "features" and "categories") represented in Autonomous decisions (Type F).

Finally, a reservation related to Automated decisions (Type F) is important to note as the implicit input–output relations—the algorithmic opacity in other words

[7]This basically describes a continuous process of "training" decision models based on previous patterns of use and/or data: it is thus also possible to envision a situation where an increased number of decisions are processed automatically (Type E) over time rather than processed manually (Type C or D) based on an explicit assessment of previous patterns of use by civil servants.

(Burrell, 2016)—do not conform well with the obligation of reason given in relation to administrative decisions which is widespread in administrative law in the Western World (Mashaw, 2007). The future will tell if empirical examples of this ideal type will increase or be limited due to this reservation.

6 Assessing the Classification's Usefulness

Following the description of the classification of the AADM usage, this section discusses its validity and usefulness as well as its practical use and a few cautions for further research.

6.1 Usefulness

Inspired by Bailey (1994) and Nickerson et al. (2013), Table 3 lists four criteria for assessing the validity and usefulness of classifications.

The first criterion—that a classification is concise, robust and exhaustive—hinges on the reasoning that configurations of decision authority between civil servants and algorithmic systems are valid for understanding AADM usage both now and in the foreseeable future. The classification conceptualises AADM usage as six different ideal typical configurations of administrative decision authority. While this is deemed appropriate in terms of both granularity and scope, new ideal types may emerge in the future due to changes in human–computer interaction, design principles and technology. For example, progress in terms of human-centred artificial intelligence (see Shneiderman, 2020) might lead to the need to expand the classification with one or more new types within its existent scope.

Table 3 Overview of criteria of validity and usefulness of classifications (building on Bailey, 1994; Nickerson et al., 2013)

Criteria	Description
Concise, robust and exhaustive	Classification must describe the phenomenon in question and do this by reducing complexity while satisfactorily grasping different variants of it.
Explanatory	Classification must "...provide explanations of the nature of the objects under study or of future objects to help us understand the objects".
Identification of similarities, differences and relationships	Classification must help identify and compare its types in relation to each other and support uncovering relationships between types.
Criteria for measurement and practical use	Ideal types of classification serve as criteria for observation and measurement thereby provide versatile and meaningful points of reference for practitioners.

The second criterion concerns the explanatory capability of classifications. At a basic level, the mapping of 10 empirical studies of AADM usage to approximate ideal type(s) in Appendix supports the validity and usefulness of the classification. A more specific aspect of the classification's strength in terms of explanation is its combinatorial power. As mentioned earlier, it is unlikely that empirical instances will exhibit the characteristics of exactly one and only one of the six types. Instead, empirical instances will exhibit combinations of the types due to the aggregated disorderliness of actual technology usage, organisational practices, agency guidance and procedures. This helps to underscore two important understandings: (i) patterns of AADM usage are ambiguous and typically involve two or more types and (ii) although five out of ten empirical examples exhibit elements of fully automated decision-making, the predominance of use seems to be the three types corresponding to semi-automated decision-support. Organisational practices entailing both Suggested procedural steps (Type C) and Automated decisions (Type E) seem particularly empirically prevalent.

Illustrative of this ambiguous pattern of use, Andersson et al. (2018) in their study of administration of driving license permits in Sweden, report that 5% of relevant cases are handled "manually" (most likely mirroring Acquisition and presentation of data (Type B)), 41% are handled semi-automatedly (mirroring either Suggested procedural steps (Type C) or Supported decisions (Type D)) and 54% are handled fully automatedly (mirroring Automated decisions (Type E)).

The third criterion describes the ability of a classification to identify and compare its types to each other, furthering the understanding of the phenomenon of AADM usage and laying the foundation for theory building. Lindgren et al. (2019) discuss the changing nature of "the public encounter" between citizens and authorities due to the digitalisation of public services. Here the classification can support a discussion of the different types and their related consequences for public encounters. Burrell (2016) and Cobbe (2019) discuss issues of transparency and opacity in relation to machine learning techniques. The classification can support discussions regarding whether problems of opacity solely "kicks in" in relation to Autonomous decisions (Type F) or relate to other types of AADM usage as well. Authors like Koulu (2020) and Peeters and Widlak (2018) have started to discuss what can be termed "algorithmic system dependency": The interlinkage of multiple systems, databases, citizen portals and network components. Here the classification can help trace how those dependencies and accompanying vulnerabilities of algorithmic systems develop in relation to the different types.

In terms of theory building, the classification can help explore patterns between types of AADM usage and wider consequences of technology use, as cases of AADM can be compared across empirical settings: are positive consequences of automation such as efficiency, increased quality and better citizen service related to specific types across empirical cases, while negative elements such as data bias, lack of transparency and "fettering" of discretion are related to other specific types? Bannister and Connolly (2020), for example, argue that the greatest risks associated with automated decision-making in the public sector are what they call "subjective/

active algorithms" which roughly corresponds to techniques employed in relation to the type of Autonomous decisions (Type F).

6.2 Practical Use

The suggested classification carries two points of reference for practical use (mirroring the assessment of the fourth criteria in Table 3). Firstly, the classification makes it possible to identify and assess ambitions of automated decision-making usage on a more informed basis. Given the complexity and criticality of a given policy area, the particular statutory regulation of the administrative decision-making and the availability and quality of data, which type of AADM usage should policy-makers and top-level managers aim for? Given such aim in terms of a type, how should the technology be tested both before deployment and during operation in order to assure satisfactory quality and use, and avoid systemic vulnerabilities across the algorithmic system?

Secondly, the classification supports more informed discussions and designs of meaningful oversight and override mechanisms (Peeters, 2020) when taking the actual use of AADM into account rather than the intended technological usage. A banal contextual factor like large case-loads of civil servants might, for example, in effect lead to organisational practices showing strong similarities to automated decisions (Type E) even though the intention might have been to support the decisions of civil servants (Type D). Based on the classification, it will be easier to assess what this and other similar discrepancies necessitate in terms of, e.g., procedures of managerial supervision and training of civils servants.

6.3 Cautions for Use

The classification also entails a few cautions for future research. Firstly, due to the primacy given to organisational practices and the function of technology, the classification is not suitable for a specific focus on predictive analytics in the public sector. Notwithstanding the particular issues related to such techniques (among others, see Gillingham, 2019; Zalnieriute et al., 2019), the classification makes us ask what the function of predictive analytics is in relation to administrative decision-making. Is alleged prediction of future behaviour of citizens or firms used as an element for the suggestion of procedural steps (Type C), in decision support (Type D) or in automated or autonomous decisions (Type E and F)? In terms of the former two types, how much emphasis (including possible over- or under-reliance) do civil servants put on suggestions?

A further caution is the exclusion of an important insight from existing classifications and typologies reviewed earlier in the chapter. The classification does not incorporate the so-called human-on-the-loop degree of automation where the human operator (the civil servant) has the ability to veto an automated decision within a certain timeframe (Parasuraman et al., 2000). This type has been left out as it has not been possible to identify any empirical instances in relation to administrative decision-making in the literature.

7 Conclusions

This chapter has focused on one general type of automation in the public sector: automated administrative decision-making (AADM). Empirically, it is widely assumed that use of AADM has increased in public administrative bodies worldwide due to technological advances. Although a number of studies have discussed the consequences of AADM both theoretically and empirically, they often only offer a simplified understanding of different uses of AADM.

Based on key references within the academic disciplines of Public Administration, Decision-support Systems and Science & Technology Studies, this chapter has conceptualised a classification of six ideal types of AADM usage. The classification maps AADM usage range from Minimal automation (Type A) to Autonomous decisions (Type F). Each type describes a configuration of decision authority between civil servants, on the one hand, and algorithmic systems on the other. While the classification of six types might be relevant to broader forms of automated decision-making at operational level (e.g. decision-making in relation to public service delivery), it should be stressed that it specifically describes instances of use of automated, *administrative* decision-making.

The suggested classification furthers the understanding of empirical AADM usage by combining focus on civil servants' technological usage with a more technical perspective allowing us to understand automated decision-making as more than a question of either being semi- or fully automated. The classification invites differentiation of broad notions of semi-automated decision-making common in much of the literature as either Acquisition and presentation of data (Type B), Suggested procedural steps (Type C) or Supported decisions (Type D) and notions of fully automated decision-making as either Automated decisions (Type E) or Autonomous decisions (Type F).

The classification gives primacy to *civils servants'* AADM usage through focus on organisational practices relating to technology rather than on technology itself. Instead of describing the intended use and "objective technology" (Fountain, 2001), the classification is a tool to map actual decision-making practices surrounding algorithmic systems. Attention must thus be paid to combinations of technology, users and cultural context. To understand AADM usage, it might be just as important to understand managerial and budgetary practices shaping the caseload of each civil servant as whether the technology is intended to facilitate Supported decisions (Type D) or Automated decisions (Type E).

In a nutshell, the classification contributes to emerging literature on automated decision-making and public service automation in three ways. Firstly, it provides a more nuanced and conceptually precise understanding of different types of AADM usage as a tool for future research. Secondly, it emphasises the importance of users' sensemaking and interpretations as well as the cultural context in order to understand the functioning and consequences of AADM. Thirdly, the classification cautions towards technologically deterministic understandings of an inevitable development towards advanced, "mature" forms of automated administrative decision-making as implied in the literature on digital government maturity and stage models (e.g. Scholta et al., 2019). Instead the classification underlines the need to understand empirical instances of use of AADM as ambiguous, often consisting of several ideal types.

For practitioners, the classification supports increased awareness of actual work-practices vis-à-vis intentions of system design in terms of AADM. The detailing of differences of types of AADM usage furthermore supports informed choices among practitioners of appropriate IT-system design and tests as well as choices of appropriate professional and management practices in relation to AADM.

Conflict of Interest The research described in this paper has been carried out as part of a PhD Project partly financed by the Danish public sector company, KOMBIT Ltd. After the author's best consideration there are no conflicts of interests.

Appendix: Empirical Examples of Ideal Types of Use of AADM

See Table 4.

Table 4 Selected, approximate empirical examples of ideal types of AADM usage. "X" indicates that the example bears resemblance to the type; and "(X)" indicates the example might bear resemblance to the type. Please note empirical examples are not based on thorough review and are simplified in descriptions. "*" indicates technology appears to no longer be in use as described

Approximate empirical example		A. Minimal automation	B. Acquisition and presentation of data	C. Suggested procedural steps	D. Supported decisions	E. Automated decisions	F. Autonomous decisions
Example	Source						
Automated administration of student grants *Education and student support, Netherlands* Technology assesses applications from students in combination with data from government databases and decides on size of grant. Civil servant has no direct role in technology operation.	Bovens and Zouridis (2002)		(X)			X	
Correctional Offender Management Profiling for Alternative Sanctions (COMPAS) *Prisons and corrections, US* Technology assesses offenders' risk of recidivism and suggests a case management plan. Civil servant decides on placement, supervision and case management of offenders in community settings prior to parole.	Skeem and Louden (2007)		X	X	(X)		

(continued)

Table 4 (continued)

Approximate empirical example		A. Minimal automation	B. Acquisition and presentation of data	C. Suggested procedural steps	D. Supported decisions	E. Automated decisions	F. Autonomous decisions
Example	Source						
Digital administration of needs of vulnerable children (DUBU) *Child welfare and protection, Denmark* Technology collects and presents selected data on cases and suggests procedural steps. Civil servant registers remaining data, assesses needs of children and decides on relevant interventions.	Fahnøe (2015)			X			
Automated re-payment of student loans *Education and student support, Sweden* Technology assesses information provided by students and decides on amount to be repaid. Casework tasks are automatically generated for civil servants in complex cases which, upon manual completion, "feeds" the technology for final decision.	Wihlborg et al. (2016)		X	X		X	

Online compliance initiative debt recovery system (OCI)* *Social security, Australia* Technology automatically identifies and notifies citizens who possibly owe money for overpayment of social security benefits. If not disproved by the citizen within a certain time frame, decision technology initiates debt recovery.	Carney (2018)					X	(X)
Automated administration of driver's licenses (Staffan) *Transport and public safety, Sweden* Technology assesses application for driving license based upon completed driving test and decides on approval or rejection. Initial rejections are automatically prepared for civil servant to assess manually.	Andersson et al. (2018)		X	X		X	
Automated administration of social security benefits *Social security, Sweden* Technology assesses applications for different social security benefits types. Simple cases are processed automatically, while complex cases or initial rejections are assessed manually by civil servant.	Ranerup and Henriksen (2019, 2020)		X			X	

(continued)

Table 4 (continued)

Approximate empirical example		A. Minimal automation	B. Acquisition and presentation of data	C. Suggested procedural steps	D. Supported decisions	E. Automated decisions	F. Autonomous decisions
Example	Source						
Automated administration of disability benefits (QDD and Insight) *Social security, US* Technology assesses applications for disability benefits, and simple cases are automatically recommended for manual decision by civil servant. Complex cases are assessed manually by civil servant. Technology flags potential errors and inconsistencies in draft decisions for manual assessment by civil servant.	Engstrom and Ho (2020)			X	X		
Automated profiling of unemployed citizens* *Public employment services, Poland* Technology assesses citizen's risk of unemployment based on information from an interview. Civil servant decides on relevant employment support within three predictive risk categories.	Kuziemski and Misuraca (2020)		X	X	X		

Automated administration of child benefits *Social security, Norway* Technology is fed by data on newborns and decides on approval or rejection of resulting application for child benefit. Applications for complex cases must be made online and are assessed manually by civil servant in combination with data from government databases.	Larsson (2021)		X			X	

References

Aghion, P., & Tirole, J. (1997). Formal and real authority in organizations. *Journal of Political Economy, 105*(1), 1–29. https://doi.org/10.1086/262063

Alter, S. (1977). A taxonomy of decision support systems. *Sloan Management Review, 19*(1), 39–56. https://doi.org/10.1016/0047-2352(80)90041-0

Andersson, A., Hedström, K., & Wihlborg, E. (2018). Automated decision-making and legitimacy in public administration. In *Scandinavian Workshop on Electronic Government (SWEG 2018)* (Issue February). http://urn.kb.se/resolve?urn=urn:nbn:se:oru:diva-73989

Averweg, U. R. (2010). Decision support systems and decision-making processes. In M. G. Hunter (Ed.), *Strategic information systems: Concepts, methodologies, tools, and applications* (pp. 122–130). Information Science Reference. https://doi.org/10.4018/9781599048437.ch025

Bader, V., & Kaiser, S. (2019). Algorithmic decision-making? The user interface and its role for human involvement in decisions supported by artificial intelligence. *Organization, 26*(5), 655–672. https://doi.org/10.1177/1350508419855714

Bailey, D. E., & Barley, S. R. (2020). Beyond design and use: How scholars should study intelligent technologies. *Information and Organization, 30*(2), 100286. https://doi.org/10.1016/j.infoandorg.2019.100286

Bailey, K. (1994). *Typologies and taxonomies: An introduction to classification techniques* (1st ed.). SAGE Publications. https://dx-doi-org.zorac.aub.aau.dk/10.4135/9781412986397

Bannister, F., & Connolly, R. (2020). Administration by algorithm: A risk management framework. *Information Polity, 25*(4), 471–490. https://doi.org/10.3233/IP-200249

Bell, J. S. (2006). Comparative administrative law. In M. Reimann & R. Zimmermann (Eds.), *The Oxford handbook of comparative law* (pp. 1259–1286). Oxford University Press. https://doi.org/10.1093/oxfordhb/9780199566020.001.0001

Bourgoin, A., Bencherki, N., & Faraj, S. (2020). "And who are you?": A performative perspective on authority in organizations. *Academy of Management Journal, 64*(Forthcoming), 1–64. https://doi.org/10.5465/amj.2017.1335

Bovens, M., & Zouridis, S. (2002). From street-level to system-level bureaucracies: How information and communication technology is transforming administrative discretion and constitutional control. *Public Administration Review, 62*(2), 174–184. https://doi.org/10.1111/0033-3352.00168

Buffat, A. (2015). Street-level bureaucracy and E-government. *Public Management Review, 17*(1), 149–161. https://doi.org/10.1080/14719037.2013.771699

Burrell, J. (2016). How the machine 'thinks': Understanding opacity in machine learning algorithms. *Big Data and Society, 3*(1), 1–12. https://doi.org/10.1177/2053951715622512

Busch, P. A., & Henriksen, H. Z. (2018). Digital discretion: A systematic literature review of ICT and street-level discretion. *Information Polity, 23*(1), 3–28. https://doi.org/10.3233/IP-170050

Carney, T. (2018). The new digital future for welfare: Debts without legal proofs or moral authority. *The Forum, University of New South Wales Law Journal, March*, 1–16. https://www.unswlawjournal.unsw.edu.au/forum_article/new-digital-future-welfare-debts-without-proofs-authority/

Cobbe, J. (2019). Administrative law and the machines of government: Judicial review of automated public-sector decision-making. *Legal Studies, 39*(4), 636–655. https://doi.org/10.1017/lst.2019.9

Cummings, M. L. (2006). Automation and accountability in decision support system interface design. *The Journal of Technology Studies, 32*(1), 23–31.

Eberle, E. J. (1984). The West German Administrative Procedure Act: A study in administrative decision making. *Dickinson Journal of International Law, 3*(1), 67–106.

Engstrom, D. F., & Ho, D. E. (2020). Algorithmic accountability in the administrative state. *Yale Journal on Regulation, 37*(3), 800–854.

Fahnøe, K. (2015). Konstruktionen af IT-systemet "Digitalisering – Udsatte Børn Og Unge" som løsningen på problemer i den socialfaglige sagsbehandling. *Nordisk Administrativt Tidsskrift, 96*(2), 37–54.
Fountain, J. E. (2001). *Building the virtual state: Information technology and institutional change.* Brookings Institution Press.
Gillingham, P. (2019). Can predictive algorithms assist decision-making in social work with children and families? *Child Abuse Review, 28*(2), 114–126. https://doi.org/10.1002/car.2547
Gilson, L. L., & Goldberg, C. B. (2015). So, what is a conceptual paper? *Group and Organization Management, 40*(2), 127–130. https://doi.org/10.1177/1059601115576425
Hauptman, A. (2013). Autonomous weapons and the law of armed conflict. *Military Law Review, 218*, 170–195.
Hogan-Doran, D. (2017). Computer says "no": Automation, algorithms and artificial intelligence in government decision-making. *Judicial Review: Selected Conference Papers: Journal of the Judicial Commission of New South Wales, 13*(3), 345–382.
Iivari, J. (2007). A paradigmatic analysis of information systems as a design science. *Scandinavian Journal of Information Systems, 19*(2), 39–64.
Jaakkola, E. (2020). Designing conceptual articles: Four approaches. *AMS Review, 10*(1–2), 18–26. https://doi.org/10.1007/s13162-020-00161-0
Kellogg, K., Valentine, M., & Christin, A. (2020). Algorithms at work: The new contested terrain of control. *Academy of Management Annals, 14*(1). https://doi.org/10.5465/annals.2018.0174
Koulu, R. (2020). Proceduralizing control and discretion: Human oversight in artificial intelligence policy. *Maastricht Journal of European and Comparative Law, 27*(6), 720–735. https://doi.org/10.1177/1023263X20978649
Kuziemski, M., & Misuraca, G. (2020). AI governance in the public sector: Three tales from the frontiers of automated decision-making in democratic settings. *Telecommunications Policy, 44*(6), 101976. https://doi.org/10.1016/j.telpol.2020.101976
Lange, A. C., Lenglet, M., & Seyfert, R. (2019). On studying algorithms ethnographically: Making sense of objects of ignorance. *Organization, 26*(4), 598–617. https://doi.org/10.1177/1350508418808230
Larsson, K. K. (2021). Digitization or equality: When government automation covers some, but not all citizens. *Government Information Quarterly, 38*(1), 101547. https://doi.org/10.1016/j.giq.2020.101547
Lindgren, I., Madsen, C. Ø., Hofmann, S., & Melin, U. (2019). Close encounters of the digital kind: A research agenda for the digitalization of public services. *Government Information Quarterly, 36*(3), 427–436. https://doi.org/10.1016/j.giq.2019.03.002
Lips, M. (2020). *Digital government: Managing public sector reform in the digital era* (1st ed.). Routledge.
Liu, C., & Graham, R. (2021). Making sense of algorithms: Relational perception of contact tracing and risk assessment during COVID-19. *Big Data & Society, 8*, 1–13. https://doi.org/10.1177/2053951721995218
Mashaw, J. L. (2007). Reasoned administration: The European Union, the United States, and the project of democratic governance. *George Washington Law Review, 76*(1), 101–125.
Nevo, S., Nevo, D., & Ein-dor, P. (2009). Thirty years of IS research: Core artifacts and academic identity. *Communications of the Association for Information Systems, 25*(1), 221–243. https://doi.org/10.17705/1CAIS.02524
Nickerson, R. C., Varshney, U., & Muntermann, J. (2013). A method for taxonomy development and its application in information systems. *European Journal of Information Systems, 22*(3), 336–359. https://doi.org/10.1057/ejis.2012.26
On-Road Automated Driving (ORAD) Committee. (2021). *Taxonomy and definitions for terms related to driving automation systems for on-road motor vehicles (J3016_202104).* https://www.sae.org/standards/content/j3016_202104/
Oswald, M. (2018). Algorithm-assisted decision-making in the public sector: Framing the issues using administrative law rules governing discretionary power. *Philosophical Transactions of the*

Royal Society A: Mathematical, Physical and Engineering Sciences, 376(2128). https://doi.org/10.1098/rsta.2017.0359

Parasuraman, R., Sheridan, T. B., & Wickens, C. D. (2000). A model for types and levels of human interaction with automation. *IEEE Transactions on Systems, Man, and Cybernetics Part A: Systems and Humans, 30*(3), 286–297. https://doi.org/10.1109/3468.844354

Peeters, R. (2020). The agency of algorithms: Understanding human-algorithm interaction in administrative decision-making. *Information Polity, 25*(4), 507–522. https://doi.org/10.3233/IP-200253

Peeters, R., & Widlak, A. (2018). The digital cage: Administrative exclusion through information architecture – The case of the Dutch civil registry's master data management system. *Government Information Quarterly, 35*(2), 175–183. https://doi.org/10.1016/j.giq.2018.02.003

Plesner, U., & Husted, E. (2020). *Digital organizing: Revisiting themes in organization studies*. Red Globe Press.

Power, D. J. (2004). Specifying an expanded framework for classifying and describing decision support systems. *Communications of the Association for Information Systems, 13*(1), 158–166. https://doi.org/10.17705/1cais.01313

Power, D. J. (2007). *A brief history of Decision Support Systems* (DSSResources.Com). http://dssresources.com/history/dsshistory.html

Ranerup, A., & Henriksen, H. Z. (2019). Value positions viewed through the lens of automated decision-making: The case of social services. *Government Information Quarterly, 36*(4), 101377.

Ranerup, A., & Henriksen, H. Z. (2020). Digital discretion: Unpacking human and technological agency in automated decision making in Sweden's social services. *Social Science Computer Review, 38*, 0894439320980434.

Riksrevisionen (S). (2020). *Automatiserat beslutsfattande i statsförvaltningen (RiR 2020: 22)*. https://www.riksrevisionen.se/rapporter/granskningsrapporter/2020/automatiserat-beslutsfattande-i-statsforvaltningen%2D%2D-effektivt-men-kontroll-och-uppfoljning-brister.html

Rosenbloom, D. H., O'Leary, R., & Chanin, J. (2010). *Public administration and law* (3rd ed.). CRC Press.

Schartum, D. W. (2018). *Digitalisering af offentlig forvaltning* (1st ed.). Fagbokforlaget.

Scholta, H., Mertens, W., Kowalkiewicz, M., & Becker, J. (2019). From one-stop shop to no-stop shop: An e-government stage model. *Government Information Quarterly, 36*(1), 11–26. https://doi.org/10.1016/j.giq.2018.11.010

Schuilenburg, M., & Peeters, R. (2021). In M. Schuilenburg & R. Peeters (Eds.), *The algorithmic society*. Routledge. https://doi-org.zorac.aub.aau.dk/10.4324/9780429261404

Seaver, N. (2019). Knowing algorithms. In J. Vertesi & D. Ribes (Eds.), *digitalSTS* (1st ed., pp. 412–422). Princeton University Press. https://doi.org/10.2307/j.ctvc77mp9.30

Sheridan, T. B. (1992). *Telerobotics, automation and human supervisory control*. MIT Press.

Shneiderman, B. (2020). Human-centered artificial intelligence: Reliable, safe & trustworthy. *International Journal of Human-Computer Interaction, 36*(6), 495–504. https://doi.org/10.1080/10447318.2020.1741118

Simon, H. A. (1960). *The new science of management decision*. Harper & Brothers Publishers.

Skeem, J., & Louden, J. E. (2007). *Assessment of evidence on the quality of the correctional offender management profiling for alternative sanctions (COMPAS)*. Center for Public Policy Research, University of California.

Smith, M. L., Martin, A. K., & Noorman, M. E. (2010). Automating the public sector and organizing accountabilities. *Communications of the Association for Information Systems, 26*, 1–16.

Stelkens, U. (2020). The impact of the Pan-European general principles of good administration in German law. In U. Stelkens & A. Andrijauskaitė (Eds.), *Good administration and the Council of Europe: Law, principles, and effectiveness* (pp. 301–329). Oxford University Press. https://doi.org/10.1093/oso/9780198861539.003.0012

Stoudt-Hansen, S., Karamouzis, F., Alexander, M., Shotton, L., Sturgill, N., & Kandaswamy, R. (2020). *Predicts 2021: Accelerate results beyond RPA to Hyperautomation* (Issue December).

Sun, T. Q., & Medaglia, R. (2019). Mapping the challenges of Artificial Intelligence in the public sector: Evidence from public healthcare. *Government Information Quarterly, 36*(2), 368–383. https://doi.org/10.1016/j.giq.2018.09.008

Veale, M., van Kleek, M., & Binns, R. (2018). Fairness and accountability design needs for algorithmic support in high-stakes public sector decision-making. In *Proceedings of Conference on Human Factors in Computing Systems (CHI '18)*, 1–14. https://doi.org/10.1145/3173574.3174014

Vogl, T. M., Seidelin, C., Ganesh, B., & Bright, J. (2020). Smart technology and the emergence of algorithmic bureaucracy: Artificial intelligence in UK local authorities. *Public Administration Review, 80*(6), 946–961. https://doi.org/10.1111/puar.13286

Weber, M. (1904/2012). The 'objectivity' of knowledge in social science and social policy. In H. H. Bruun & S. Whimster (Eds.), *Max Weber: Collected methodological writings* (pp. 100–138). Routledge.

Widlak, A., van Eck, M., & Peeters, R. (2021). Towards principles of good digital administration: Fairness, accountability and proportionality in automated decision-making. In M. Schuilenburg & R. Peeters (Eds.), *The algorithmic society: Technology, power, and knowledge* (pp. 67–83). Routledge.

Wihlborg, E., Larsson, H., & Hedström, K. (2016). "The computer says no!" - A case study on automated decision-making in public authorities. *Proceedings of the Annual Hawaii International Conference on System Sciences, 49*, 2903–2912. https://doi.org/10.1109/HICSS.2016.364

Zalnieriute, M., Moses, L. B., & Williams, G. (2019). The rule of law and automation of government decision-making. *Modern Law Review, 82*(3), 425–455. https://doi.org/10.1111/1468-2230.12412

Zouridis, S., van Eck, M., & Bovens, M. (2020). Digital discretion. In T. Evans & P. Hupe (Eds.), *Discretion and the quest for controlled freedom* (pp. 313–329). Palgrave Macmillan. https://doi.org/10.1007/978-3-030-19566-3

Cognitive Robotic Process Automation: Concept and Impact on Dynamic IT Capabilities in Public Organizations

Gustaf Juell-Skielse, Prasanna Balasuriya, Evrim Oya Güner, and Shengnan Han

1 Introduction

Robotic Process Automation (RPA) is becoming a significant aspect of modernizing and digitally transforming public sector organizations (Houy et al., 2019). RPA is a process automation technology used for reducing repetitive and manual labor in administrative and standardized tasks where software robots perform work previously done by humans (Willcocks et al., 2015). The software robots are designed to imitate human actions in tasks and work processes by interacting with information systems through existing user interfaces (Lacity & Willcocks, 2016; Penttinen et al., 2018; Dias et al., 2019).

The research about RPA in the public sector consists, with a few exceptions, of case studies of municipalities and universities. The available research increases the understanding of the implications of using RPA to automate different processes for social services (Nauwerck & Cajander, 2019; Ranerup, 2020; Ranerup & Henriksen, 2019, 2020), case handling in general (Lindgren, 2020) and administrative internal processes, such as human resources (Denagama Vitharanage et al., 2020; Eikebrokk & Olsen, 2020; Goday-Verdaguer et al., 2020, Patil et al., 2019).

Research on RPA in public sector organizations has primarily been delimited to tasks based on structured information, e.g., Patil et al.'s automated analysis of exam results (Patil et al., 2019). However, public organizations encounter many situations where decisions and actions are based on unstructured or semi-structured data, e.g., Goday-Verdaguer et al., (2020) investigate how RPA can use unstructured data from

G. Juell-Skielse (✉)
University of Borås, Borås, Sweden
e-mail: gustaf.juell-skielse@hb.se

P. Balasuriya · E. O. Güner · S. Han
Stockholm University, Stockholm, Sweden
e-mail: balasuriya@dsv.su.se; evrim.guner@dsv.su.se; shengnan@dsv.su.se

G. Juell-Skielse et al. (eds.), *Service Automation in the Public Sector*, Progress in IS,
https://doi.org/10.1007/978-3-030-92644-1_4

emails to improve hiring processes, and Houy et al. (2019) develop a software robot that can capture and process unstructured information from tax notices. In these situations, RPA can become more useful if it includes functions based on artificial intelligence for capturing and processing unstructured and semi-structured data, such as, machine learning and natural language processing. Integrating RPA with different AI-technologies, so-called *cognitive RPA* (Viehhauser, 2020; Houy et al., 2019; Suri et al., 2018), could increase the adoption of RPA as it extends its applicability to solve more complex tasks (van der Aalst et al., 2018). However, we lack a general understanding of the concept of cognitive RPA.

Due to the technical advances of RPA, the use of RPA is expected to generate IT-related capabilities in public organizations. So far, judged by available research concerning RPA in public organizations, RPA is considered as an IT resource primarily used for reducing repetitive and manual tasks based on structured data. However, we lack a sufficient understanding of how RPA, as a type of IT resource, affects public organizations' dynamic IT capabilities (Li & Chan, 2019), i.e., how public organizations can change their IT-related capabilities to adapt to new situations by utilizing RPA.

This chapter is aimed to mitigate two research gaps: (1) a lack of conceptualization of cognitive RPA and (2) a lack of understanding of the impacts of RPA on public organizations' dynamic IT capabilities. The research data were collected from 13 interviews with seven RPA vendors. We adopted an abductive approach in the thematic analysis of interview data. As a result, we define and conceptualize *cognitive RPA* and propose a set of propositions for how RPA affects public sector organizations' dynamic IT capabilities.

The chapter is structured accordingly. Section 2 presents robotic process automation and dynamic IT capabilities. Section 3 contains the methodological choices and Sect. 4 the results. In Sect. 5, the findings are discussed, followed by conclusions and future research in Sect. 6.

2 Extended Background

The extended background introduces research on robotic process automation and dynamic IT capabilities.

2.1 Robotic Process Automation

RPA technology is used to automate structured processes by executing repetitive, manual, rule-based, and high-volume routine tasks. Willcocks et al. (2015) define RPA as the automation of service tasks by configuring software that controls robots to perform the work previously done by humans. Similarly, Penttinen et al. (2018) and Dias et al. (2019) describe RPA as a process automation technology configured to perform work by interacting with information systems through existing user interfaces (Lacity & Willcocks, 2016; Penttinen et al., 2018; Dias et al., 2019). The configured software is often called a "software robot" that is instructed and assigned to execute tasks (van der Aalst et al., 2018) as human users do in process-aware systems (Syed et al., 2020).

The concept of robot denotes a human-like machine imitating certain human movements and functions automatically.[1] A software robot is not human like in the physical sense, rather its focus is on imitating human actions when working with information systems to perform work processes automatically. By automation we here refer to the automatic execution of tasks included in a work process by a computer program with little or no direct human control. Hence, software robots are computer programs with functions that imitate human actions on computer interfaces to automatically perform work processes.

Robotic process automation replaces people by working from outside-in without changing the information system (van der Aalst et al., 2018). This also adds to the notion of a software robot that uses an information system rather than being an integral part of it and therefore change the information system design from within. Being non-invasive to the underlying information systems, RPA easily conforms with the enterprise safety requirements, i.e., security, scalability, and auditability (Stolpe et al., 2017).

2.2 *RPA Vendor Platforms*

To set up and deploy software robots, RPA vendors offer standardized software platforms to users of RPA (Willcocks et al., 2015). Two of the major RPA software vendors are Uipath and Automation Anywhere (Ray et al., 2020). RPA software platforms typically include tools for developing and operating software robots. For example, Uipath's RPA platform includes a developer studio used to design and configure software robots and an orchestrator used to deploy, operate, and monitor software robots in the user's organization. Similarly, Automation Anywhere's RPA platform includes a bot creator, a bot runner and an enterprise control room. Also, RPA vendors offer tools to utilize AI technology in software robots. For example, Uipath offers the AI Fabric to deploy AI-models used by software robots. Similarly, Automation Anywhere offers the IQ bot portal to create and train software robots to process semi-structured and unstructured documents. AI-models, deployed through vendors' platforms, are trained for the tasks that the software robots are executing. Also, exceptions handled through human interventions could be used for improving the AI-models.

Currently, the functions of cognitive RPA are often not included in vendors software packages. Viehhauser (2020) argues that it is not reasonable to expect

[1] https://www.lexico.com/definition/robot.

Table 1 A framework for classifying cognitive intelligence (based on Viehhauser, 2020, p. 104)

Capture Information
– Text and character recognition
– Image recognition
– Voice and sound recognition
Process Information
– Natural language processing
– Machine learning
– Automated reasoning

these functions to be built-in to RPA vendors' software since they are complex and developing at a high pace. Instead, RPA vendors rely on services provided by partners but simplify the integration of these services through their platforms.

2.3 *Cognitive RPA*

Artificial intelligence could be defined as "a system's ability to correctly interpret external data, to learn from such data, and to use those learnings to achieve specific goals and tasks through flexible adaptation" (Kaplan & Haenlein, 2019, p. 15). The combination of RPA and AI-technologies is sometimes referred to as *cognitive RPA* (Viehhauser, 2020; Houy et al., 2019; Suri et al., 2018) as it utilizes a type of artificial intelligence for emulating cognitive intelligence, i.e., cognitive computing for pattern recognition and systematic thinking (Viehhauser, 2020). Cognitive computing is based on several AI-technologies, such as machine learning, neural networks, natural language processing and automated reasoning (Davenport & Kirby, 2016). By combining RPA with artificial intelligence for cognitive computing, the software robot increases its ability to capture and process information. By doing so, RPA extends towards the realm of cognitive automation, moving from rules to inferences (Lacity & Willcocks, 2016). Viehhauser (2020) presents a framework for classifying the resulting cognitive intelligence, see Table 1.

Available research on cognitive RPA in public sector organizations includes two cases, one about process analysis (Goday-Verdaguer et al., 2020) and one concerning document processing (Houy et al., 2019). Goday-Verdaguer et al. (2020) present a case study of a municipal hiring process. The municipality needed to improve its hiring process to increase quality and to reduce costs. The authors used a combination of RPA and process analysis. Unstructured information from emails was captured and analyzed using process mining methods to visualize the municipal hiring process and to evaluate its performance. Houy et al. (2019) present a use case where companies want to automatically determine taxes when trading in several municipalities. However, trade tax assessment notices are often paper documents and come in various layouts, depending on municipality. The authors present a cognitive RPA prototype that automatically captures unstructured data from trade tax notices, with variations in layouts, which are then processed and automatically entered into companies' accounting systems.

Other available studies about RPA in public sector organizations concern applications using structured data. They are primarily based on case studies of municipalities and universities using RPA to automate work processes and decision-making. Ranerup and Henriksen (2020) examine the effect on local civil servants' discretionary practices of the use of RPA. Ranerup and Henriksen (2019), reporting from a case from municipal social services, found that the adoption of RPA contributed to several public sector value positions: professionalism, efficiency, and service. Nauwerck and Cajander (2019) study the benefits and challenges for caseworkers in social services. Ranerup (2020) investigates the use of RPA to enable aspirational changes in decision-making in social assistance applications, and Lindgren (2020) presents an ongoing project to develop a tool for selecting case handling processes to automate. Patil et al. (2019) analyze the benefits of using RPA for the analysis of exam results, and Denagama Vitharanage et al. (2020) contribute with a framework of benefits for using RPA in a university payroll process. Dias et al. (2019) present a case study of a national government service center and aim to explain knowledge embodiment in knowledge work being performed by humans and machines together.

The lack of studies about cognitive RPA in public sector organizations means that the understanding of utilizing cognitive RPA for handling tasks based on unstructured data is still limited. Also, none of the presented studies discuss how the dynamic capabilities of the public organizations are affected by adding RPA as an IT resource. Therefore, we see a clear gap in the literature for understanding how RPA can be applied to situations where public organizations make decisions based on semi- and unstructured data and how the use of RPA, including cognitive RPA, affects public organizations' IT-related capabilities.

2.4 *Dynamic IT Capabilities*

To study how the use of RPA affects IT-related capabilities in public organizations, we adopt the perspective of dynamic capabilities (Teece et al., 1997). Dynamic capabilities are "...organizational and strategic routines by which firms achieve new resource configurations..." (Eisenhardt & Martin, 2000, p. 1106). Ordinary capabilities are patterned and repetitive processes and routines that make up the operations of an organization. Dynamic capabilities are routines and processes that create, extend and modify ordinary capabilities when ordinary capabilities are insufficient to adapt to new situations.

Public sector organizations exist to serve the needs of citizens or particular client groups, rather than making profits (Collins, 2005). Therefore, strategies and plans tend to focus on maximizing organizational performance by building on internal resources and key actors (Pablo et al., 2007). According to Pablo et al. (2007), managers in public organizations identify latent dynamic capabilities, build on established levels of trust to use these dynamic capabilities, while balancing the tension between local development and central control.

To increase the understanding of the role of IT in the development of dynamic capabilities, Li and Chan (2019) introduced dynamic IT capabilities. Li and Chan build on a combination of dynamic capabilities, and Bharadwaj's (2000) classification of IT resources when they identify three components of dynamic IT capabilities: (1) dynamic digital platform capability, (2) dynamic IT management capability and (3) dynamic IT knowledge management capability. Also, they distinguish between ordinary IT-related capabilities and first-order dynamic IT capabilities. Ordinary IT-related capabilities are associated with deployment and support of IT resources, such as setup and implementation of a sales system, while dynamic IT capabilities are developed to create, modify, and extend ordinary IT capabilities, such as methods for setting up an enterprise system. Li and Chan (2019) argue that these first-order dynamic IT capabilities are key to organizations' ability to transform digitally. However, given the available research about RPA and cognitive RPA in public organizations presented in Sect. 2.3, it is unclear how RPA, including cognitive RPA, influences public organizations' dynamic IT capabilities.

3 Method

To conceptualize a capability framework of RPA from the RPA suppliers' perspective, this study adopted a qualitative explorative approach (Myers & Avison, 2002). The study was explorative because we have limited knowledge of what the concept of cognitive RPA is and the impacts of it on public organizations' dynamic IT capabilities. Qualitative methods are appropriate for an explorative study. Moreover, the study followed the principles of conceptual research in information systems discipline (Mora et al., 2008). Guided by the conceptual research, the study aims to develop and justify the conceptual framework (or construct) for understanding cognitive RPA and its effects on IT capabilities in public organizations.

The research data were collected in three steps. First, we did a Google search of all RPA system suppliers and cross-checked with the suppliers' list from Ray et al. (2020). We identified 16 RPA suppliers. Second, we searched the contact information of these suppliers and sent formal emails to request interviews. After sending several reminders, we received positive supports from seven RPA suppliers, which are Automation Anywhere (USA), Blue Prism (Sweden), JOLT Advantage Group (USA), SAP (Sweden), PS provider (Sweden), Robusta Cognitive Automation (Turkey), UiPath (Romania, Sweden, USA). Third, we developed a semi-structured interview protocol including both retrospective and prospective questions (Schultze & Avital, 2011). The 13 interviews were conducted from May to October 2020, see Table 2. All interviews were facilitated by either Zoom or Microsoft Teams. The interviewees' names and their associated working companies are kept anonymous. Table 2 shows an overview of the conducted interviews. In qualitative information systems research, interview is a common method for gathering data and is an appropriate method to gather explorative information regarding the effects of information systems on organizations (Myers & Avison, 2002). At the end, we used a

Table 2 Overview of respondents, working positions, interview dates and lengths

Respondent	Respondent's working position	Length	Date
Respondent 01 (R1)	Senior Vice President of Operations	85 min	30 September 2020
Respondent 02 (R2)	Vice President, Chief Evangelist	48 min	06 October 2020
Respondent 03 (R3)	CTO, Director of Public Sector Marketing	60 min	06 October 2020
Respondent 04 (R4)	Director Solution Management Intelligent RPA, New Ventures and Technologies	52 min	09 October 2020
Respondent 05 (R5)	Head of Presales & Customer Success, Public Sector	68 min	14 October 2020
Respondent 06 (R6)	Customer Success Director	55 min	15 October 2020
Respondent 07 (R7)	Co-Founder	43 min	15 October 2020
Respondent 08 (R8)	Senior Presales Engineer	52 min	15 October 2020
Respondent 09 (R9)	Vice President—Public Sector Industry	87 min	19 October 2020
Respondent 10 (R10) Respondent 11 (R11)	Executive Vice President of Products Senior Vice President of Products	49 min	21 October 2020
Respondent 12 (R12)	Machine Intelligence Automation Consultant and data scientist	91 min	22 October 2020
Respondent 13 (R13)	RPA Consultant	95 min.	20 May 2020

combination of manual transcription and the Trint[2] software service to transcribe the interview data. The resulting transcription document includes approximately 61,000 words.

The collected data were thematically analyzed by applying the step-by-step guidelines provided by Braun and Clarke (2006), which are: familiarization with data, generating initial codes, searching for themes among codes, reviewing themes, defining and naming themes, and producing the final report. This technique brings flexibility to researchers to identify the themes as well as requires rigor and consistent analysis of the data (Guest et al., 2011). Three authors were involved in the first steps of the thematic analysis. The team did the last three steps jointly to ensure reliability and validity of the results. The final themes were achieved through discussions and working meetings among the team. Moreover, we adopted an abductive approach in the thematic analysis (Fereday & Muir-Cochrane, 2006), which enabled us to combine the inductive and deductive thinking in conceptualizing cognitive RPA and proposing the IT capability framework (Dubois & Gadde,

[2] https://trint.com/.

2002; Mora et al., 2008). By doing so, we were able to develop the concept of cognitive RPA and the IT capability framework with both sound theoretical underpinnings and the themes identified from the interview data. This also aided us to deductively analyze the impacts of RPA on IT-related capabilities in public organizations. However, we shall take a critical stance of using RPA suppliers' data in analyzing the effects of cognitive RPA on the IT capability in public organization. The suppliers' perspectives and reflections may tend to overestimate the effects (Kallinikos, 2004). Nonetheless, the system characteristics, the functional logic and overall capabilities of cognitive RPA as we conceptualize in this chapter are equally applicable to public organizations.

4 Results

We conceptualize cognitive RPA, including its intelligent capabilities for document processing, image processing, conversation processing and process analysis. We further analyze the impacts of RPA, including cognitive RPA, on public organizations' dynamic IT capabilities from three aspects: (1) dynamic digital platform capability, (2) dynamic IT management capability and (3) dynamic IT knowledge management capability.

4.1 Unpacking Cognitive RPA: System Capabilities

The general view among the respondents is similar to literature and they consider that cognitive RPA is a combination of RPA and AI-technologies, where AI-technologies extend the scope of decisions that can be automated while RPA automates actions based on these decisions. R8 believes that cognitive RPA makes the software robots more intelligent and gives them more capabilities so that they can mimic human behavior in a better fashion. R12 believes that cognitive RPA makes AI more actionable as "AI alone is just, you know, a decision." R5 views cognitive RPA as a key to make AI-technologies implementable in a wide range of business situations. Or as R8 puts it "[if] a customer [is] already working with IBM Watson, then we have . . . an integration where our robots can utilize and take the benefit of using Watson." R4 sees specific application areas and certain scenarios where cognitive RPA is useful and based on combinations with particular AI-technologies, e.g., machine learning and natural language processing.

Thus, we define *cognitive RPA* as a category of robotic process automation (RPA) where software robots have the intelligent capabilities to extract, interpret, and act, mimicking human behavior when using information systems, on structured and unstructured data to automate work processes. In particular, cognitive RPA possesses intelligent capabilities for document processing, image processing, conversation processing and process analysis.

Based on this definition, RPA, including cognitive RPA, can be conceptualized as an open system, comprising of input, process, output, and feedback, see Fig. 1. Structured and unstructured data is captured and processed using rules and cognitive artificial intelligence to produce useful and meaningful information. Feedback is used to improve RPA performance. It is referred to as an open system (von Bertalanffy, 1968) to emphasize the exchange of information with the environment and the feedback loops. Compared to Viehhauser's (2020) framework, presented in Fig. 1, the open system in Fig. 1, makes visible the interactions with the environment, input and output, as well as the feedback loop for handling, e.g., exceptions. The input interactions with the environment are implicit in Viehhauser's framework through the concept of capturing information. However, output and feedback are not that clear.

4.1.1 Document Processing

Based on the interview data from respondents R2, R3, R5, R6, R8, R9, R11, and R12, we summarize the RPA capability of document processing. Document processing means that a software robot has the capacity to read, interpret, extract as well as act upon unstructured and semi-structured text data. AI-technologies, such as natural language processing, are used to automatically determine the location of data in documents with variations in structure and layout, such as emails, bills, chats, and receipts. Standard RPA can be used for document understanding as well, however, it requires that the documents include structured data, such as licenses, passports or forms.

Document understanding involves the creation of taxonomies to categorize data of a document and to enable the extraction of information that a software robot can act upon. The taxonomy is continuously developed as the model is exposed to new instances and feedback from human validation. Document understanding is often combined with OCR technology (Optical Character Recognition) to capture text from written and typed text and convert it to machine-readable formats. For example, software robots with document understanding can capture text from emails and classify the emails with the help of natural language processing. Based on the classification, the software robot can take different actions, such as routing the email to a certain user or fill out a form based on the unstructured information provided in the document. Moreover, software robots with document understanding can interpret bills from suppliers, with various layouts and structures, and automatically book the bills in the book-keeping system.

Document processing could be described as a sequence of steps. First, a taxonomy is chosen by the software robot. Second, the text is captured and digitized. Third, the text is classified using natural language processing. Fourth, information is extracted from the document. Fifth, the extracted information is exported for further use. In supervised models, the extracted information can be validated by humans and changes made by humans can be used to improve the taxonomy.

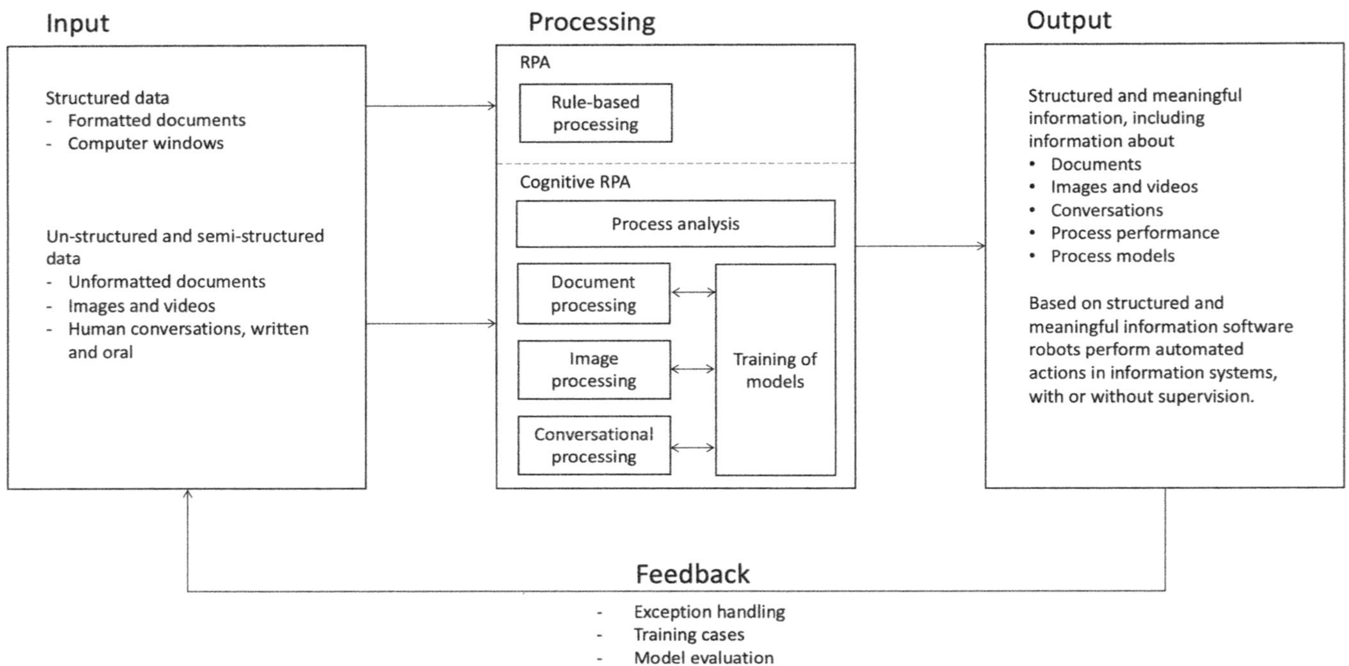

Fig. 1 A general systems view of RPA, including cognitive RPA

4.1.2 Image Processing

Based on the interview data from respondents R1, R2, R5, R6, R7, R9, R10, and R12, we summarize the RPA capability of image processing. Image processing means that a software robot has the capability to capture and process information from images, videos, and other visual inputs. Common applications are to process screen elements of user interfaces as images containing unstructured data and to handle variations in layouts of paper documents, such as bills and receipts.

Image processing often uses optical character recognition (OCR) and artificial intelligence, such as computer vision and machine learning. OCR is a set of techniques utilized for the conversion of characters in graphic files to machine-readable formats. Computer vision includes techniques, such as convolutional neural networks and deep learning, to segment images, classify images, detect and track objects and make decisions.

Image processing could be described as a sequence of steps. First, semi-structured and unstructured data, e.g., images, is captured. Second, the data is recognized as an image, using object detection and shape analysis. Third, the image is analyzed by using computer vision and deep learning techniques for segmentation and classification. If needed, the image is reconstruct using techniques for image enhancement, noise removal and feature detection. Fourth, the result of the analysis is used to automatically make a prediction or decision and used by the software robot for further actions. Fifth, exceptions handled by humans can be fed back as input to retrain the models used for analysis.

The models used for image processing need to be trained for the task performed by the software robot. Normally there is an initial data set available for an image processing model but it has to retrained to be adapted to the specific task at hand. To train the model, relevant features are extracted from the images or videos to make a meaningful interpretation of the data. Deep learning techniques are often used for this purpose. Also, convolutional neural networks, a deep learning technique, are used for classifying image content.

When image processing is used by software robots to interpret user interfaces, the trained model automatically detects various types of screen elements such as textboxes, dialog windows, and click boxes. This is useful in situations when software applications reside in a remote or virtual environment (Martins et al., 2020) when user interfaces are not presented as structured information.

Public sector application of image processing includes, e.g., automatic processing of receipts of various layouts. Employees can send photos of receipts to image processing robots that scan and analyze images of the receipts and automatically register them in an accounting system. Another example is the US State department's use of image processing to validate passports. Signatures and photos of passports are handled by image processing robots using OCR and computer vision techniques to quickly make predictions with high levels of confidence.

Similar to image processing, video processing is also becoming used by public organizations. As the AI-models for video processing mature, software robots with an ability of capturing and processing information from videos accurately will become more common.

4.1.3 Conversation Processing

Based on the interview data from respondents R1, R2, R3, R4, R5, R6, R9, R10, R11, and R12, we summarize the RPA capability of document processing. Conversation processing provides functionality to an RPA software robot to handle conversations with human actors. An example of conversation processing is when a person contacts a citizen service center and describes a need, using speech, and a software robot can serve the person to satisfy that need. For example, by providing a form and explaining how it should be used. In other words, conversation processing means that a software robot can include functions similar to a chatbot (Dahiya, 2017).

> So, [a citizen] asks a question to the chatbot. The chatbot sends it to the [RPA]-robot that captures information from the systems, and then the information is presented to [the citizen]. (Respondent R6).

When configuring a software robot for conversation processing, a conversation model is added to the workflow of the software robot, such as Google Dialogflow[3] or Druid AI.[4] It is possible to add conversation models that are pre-configured for certain domains, such human resources and sales, and certain industries, such as banking and healthcare.

Conversation processing includes several steps: capturing human utterances, identifying intents and extracting entities (Singh et al., 2019). Based on the intent, the software robot performs actions. When the intent cannot be identified with confidence during a conversation, there is a fallback procedure for human assistance. Also, the result of human assistance provides training data of the conversation model. Conversation processing utilizes technology for natural language processing, which includes, e.g., classification of text, tokenization of sentences, stemming of words, tagging words into part-of-speech, parsing text and semantic reasoning.

> We make use of natural language processing when we combine chatbots and RPA robots for certain scenarios. (Respondent R4).

4.1.4 Process Analysis

Process analysis is the initial step in the RPA lifecycle (Jimenez-Ramirez et al., 2019). It provides a view of the real process flow and an understanding of how the

[3] https://cloud.google.com/dialogflow.

[4] https://www.druidai.com.

process operates. Process analysis allows to improve the processes by identifying the roots of the inefficiencies, e.g., duplications, delays, and deviations.

A common method used in process analysis is process mining. Process mining refers to a technique of data-driven process analysis to visualize the flow of business processes based on the event logs extracted from different data sources, e.g., enterprise resource planning systems (Geyer-Klingeberg et al., 2018). By using the event logs, three main types of process mining are performed: process discovery e.g., extracting process models; conformance checking, e.g., monitoring deviations by comparing model and log; enhancement, e.g., extending or improving an existing process model (Van Der Aalst, 2012).

Through process mining, manual approaches to routine identification, e.g., interviews with process owners or observations, are automated (Geyer-Klingeberg et al., 2018; Agostinelli et al., 2019), the undesired process patterns and the deviations from the ideal process are identified (Geyer-Klingeberg et al., 2018) and the software robots are monitored (van der Aalst, 2020). This helps to streamline the processes not only prior to an RPA initiative (Jimenez-Ramirez et al., 2019; Geyer-Klingeberg et al., 2018) but also after the automation (van der Aalst., 2020) for further governance and enhancement of those processes (Kirchmer & Franz, 2019).

Artificial intelligence helps to identify the processes with the maximum automation potential to accelerate the implementation of RPA and thus decreasing the cost of the robots (van der Aalst et al., 2018).

> It is an AI engine that can watch the process, determine how many steps are in the process, the average time of the process, the number of times you go for approval, sitting in queues, those types of things, and then, make recommendations as how you change the process to make the business better and more efficient. (Respondent R9).

RPA initiatives are proved to be more efficient when the automation is incorporated with AI-technologies. The case of Vodafone illustrates that the implementation of RPA combined with process mining improved the transparency of the business processes and eased the identification and prioritization of the automation alternatives (Lazarus, 2018; Geyer-Klingeberg et al., 2018). Similarly, Goday-Verdaguer et al. (2020) report how process mining is utilized in the identification of the automation potentials in a hiring process. In line with prior research, R5 believes that the public sector would benefit from RPA with cognitive capabilities for process analysis to streamline their internal processes, increase efficiency and deliver more value to the citizens.

4.2 *RPA as Dynamic IT Capability*

In this results section, we analyze how the extended notion of RPA, including cognitive RPA, affects public organizations' dynamic IT capabilities. The result of the analysis is presented from the perspective of the three components introduced by Li and Chan (2019): (1) dynamic digital platform capability, (2) dynamic IT management capability and (3) dynamic IT knowledge management capability. In turn, these three dynamic IT capabilities affect nine IT-related ordinary capabilities (ibid.), which is further discussed below.

4.2.1 Dynamic Digital Platform Capability

Dynamic digital platform capability enables organizations to create value from their digital platforms and influences three IT-related ordinary capabilities: IT infrastructure capability, IT integration capability and IT infrastructure flexibility.

IT infrastructure capability is the capacity of a public organization's IT infrastructure to support its activities and processes (Li & Chan, 2019). Based on the interview data from respondents R3, R5, R7, R11, and R 12, we summarize the impact of RPA on IT infrastructure capability. RPA adds rule-based automation on structured data to the IT infrastructure capability by imitating how humans work on information systems' user interfaces.

> RPA three, four or five years ago was very simplistic. It was rules-based. It simply mimicked the behavior of humans as they move their mouse or do a keystroke. (Respondent R3).

This means that the IT infrastructure can be used to automate more activities and processes. Vendors of RPA software provide platforms and cloud services for setup, deployment and maintenance of software robots. Also, cognitive RPA provides infrastructure for artificial intelligence to process unstructured and semi-structured data to further automate activities and processes.

> Cognitive RPA would be the key to enabling AI technologies and smart technologies across the board. (Respondent 5).

As cognitive RPA with unstructured data works on interfaces of information systems, it provides a relatively easy way to integrate artificial intelligence in IT-enabled processes.

IT integration capability refers to the capacity of a public organization to use its IT infrastructure for intra- and inter-organizational coordination (Li & Chan, 2019). Based on the interview data from respondents R1, R2, R4, R5, and R8, we summarize the impact of RPA on IT integration capability. RPA adds to the IT integration capability by enabling automatic data transfer between information systems imitating human actions on the interfaces.

> [RPA] might not be the fanciest approach [to systems integration]. Often, it's used as no proper integration is available, but in the end, it helps many customers and generates business. (Respondent R4).

Also, RPA platforms (or services) provide a technical environment for setup, deployment and scaling of software robots. It adds to the IT integration capability by simplifying technical integration and shortening the lead times for implementing technical integrations. Also, cognitive RPA adds to the IT integration capability through artificial intelligence. It can facilitate the automation of tasks that involve judgment where cognitive RPA mimics human thought processes.

When you start to apply cognitive reasoning to RPA it can now make a judgement call. (Respondent R2).

Also, through the use of document understanding and adjacent AI-technologies, cognitive RPA adds to the IT integration capability to communicate more effectively with external organizations, for example, in a value chain.

Cognitive RPA, with AI built into the bots, themselves are innovations with other capabilities..., it could be NLP, it could be machine learning, ... anything else that actually brings that end-to-end solution that the customer is looking for. (Respondent R5).

IT infrastructure flexibility refers to the extent a public organization's existing IT infrastructure can be adapted to meet new objectives and requirements (Li & Chan, 2019). Based on the interview data from respondents R9 and R12, we summarize the impact of RPA on IT infrastructure flexibility. RPA adds to the infrastructure flexibility through its scalability and short implementation times and through its integrative functionality. Additionally, RPA is available as cloud services and can be scaled up and down due to needs. Moreover, public organizations can share automation flows through vendors' repositories. Also, cognitive RPA adds to the infrastructure flexibility by lowering error rates through reduced human involvement. Moreover, cognitive RPA is a way to extend the IT infrastructure flexibility by adding artificial intelligence. Through cognitive RPA, it becomes easier for organizations to deploy and create value from artificial intelligence.

You can deploy any kind of AI machine learning solution that you want on top of this [RPA] platform. So now the door is open to any kind of solution. (Respondent R12).

4.2.2 Dynamic IT Management Capability

Dynamic IT management capability refers to the capacity of a public organization to design and change processes that control IT resources in response to changing goals and priorities. It influences three IT-related ordinary capabilities: IT deployment capability, IT exploitation capability and IT exploration capability.

IT deployment capability refers to the capacity to assist employees to adopt, use, and retire IT resources (Li & Chan, 2019). Based on the interview data from respondents R1, R9, and R11, we summarize the impact of RPA on IT deployment capability. RPA deployment is often governed by centers of excellence in a federated model. Employees and citizen developers are guided by the centers of excellence to set up and deploy software robots (Forrester, 2014; Anagnoste, 2018). The center of excellence can enhance and add complexity to the automations to make them more versatile and scalable throughout the entire organization.

...enabling their citizen developers or workers to build the automation governed by the Center of Excellence in a federated model. (Respondent R1).

Moreover, software robots developed to handle peak situations can easily be terminated when no longer needed. Also, RPA adds to the IT deployment capability through re-thinking and re-designing the IT deployment methods, such as embracing agile approaches in RPA projects. Additionally, cognitive RPA can, through process understanding, learn about business processes to increase automation and to make available software robots more scalable and versatile.

> So let the process run and then have . . . an AI engine that can watch the process and . . . make recommendations as to how you change the process to make the business better and more efficient. (Respondent R9).

IT exploitation capability is the capacity to effectively use existing IT resources to support activities and processes (Li & Chan, 2019). Based on the interview data from respondents R1, R2, and R5, we summarize the impact of RPA on IT exploitation capability. RPA adds to the IT exploitation capability with its functionality for automation through software robots. Software robots can be set up for automatically handling a variety of activities and processes that are difficult or more expensive to develop using traditional systems engineering. Also, the aggregate use of several, yet simple software robots, can have a significant impact on the work performed in an organization.

> It's a pretty simple concept when they are by themselves. But actually, when combined, they make a very powerful solution for customers. (Respondent R1).

RPA adds to the IT exploitation capability by its cost-efficient nature. It is relatively simple for citizens and civil servants to set up and use software robots for different situations. Also, cognitive RPA provides an environment for rapid and simple exploitation of artificial intelligence to develop new and enhance existing business processes within and between organizations.

> The whole idea of [cognitive RPA] is to take those predictions or those prescriptions created by the machine learning model or the algorithm and then give this ability to act on what RPA does. (Respondent R2).

IT exploration capability is the capacity to experiment and discover innovations in IT resources and practices (Li & Chan, 2019). Based on the interview data from respondents R1, R3, and R5, we summarize the impact of RPA on IT exploitation capability. RPA adds to the IT exploration capability by making automation more accessible and affordable for public organizations. Federated models, governed by centers of excellence, involving civil servants, citizen developers and IT staff support experimentation and discovery of automated solutions. Also, cognitive RPA adds to the IT exploration capability through its ability of automating non-standard processes, which requires artificial intelligence. Some vendors refer to this as a "digital workforce" which provide different skills than the human workforce and at a different cost.

> And just like in the human workspace where we have different people with different skills and different salaries, I think you're going to bring in this [digital workforce], all of these digital employees, and ask them to do what they do best. (Respondent R3)

Moreover, cognitive RPA adds to the IT exploration capability by expanding the public organizations' partner network to also include AI technology providers.

4.2.3 Dynamic IT Knowledge Management Capability

Dynamic IT knowledge management capability is a public organization's capacity to create, transfer and retain organization-wide IT-related knowledge. It influences three first-order capabilities: IT knowledge creation capability, IT knowledge transfer capability and IT knowledge retention capability.

IT knowledge creation capability is the capacity to create useful IT knowledge for the organization (Li & Chan, 2019). Based on the interview data from respondents R1, R3, and R5, we summarize the impact of RPA on IT knowledge creation capability. RPA affects the IT knowledge creation through its ability to automate knowledge discovery and through federated deployment models. RPA can be used both to automate process analysis, see Sect. 4.1.4, and to automate data mining. A recent example of how data mining in organizations can be automated using software robots is the company Datarobot[5] which provides robots that automatically captures data and process this data, using artificial intelligence. Dias et al. (2019) investigate knowledge creation in situations where humans and software robots interact and uncover several stages of knowledge embodiment, e.g., cognitive reasoning and cognitive scaffolding. Also, the federated deployment model supports knowledge creation by, e.g., establishing feedback loops between local applications and centers of excellence. Also, the lightweight IT characteristics of RPA simplify for business people to engage in development of IT-related solutions and thus in IT knowledge creation. Also, cognitive RPA adds to the IT knowledge creation as the organizations become capable of turning their RPA robots into smarter robots that are able to process a wide variety of unstructured content. RPA, as lightweight IT and integrator of artificial intelligence, becomes a vehicle for creating IT knowledge for utilizing AI-technologies in value creation.

IT knowledge transfer capability is the capacity to make useful IT knowledge available in a timely manner to relevant stakeholders within and outside the organization (Li & Chan, 2019). Based on the interview data from respondents R1, R3 R5, and R9, we summarize the impact of RPA on IT knowledge transfer capability. RPA vendors provide different resources, such as forums, to share knowledge about RPA applications. By developing centers of excellence with a mixture of business and IT knowledge related to RPA, organizations can increase their ability to disseminate knowledge about situated RPA applications. Also, cognitive RPA adds to the IT knowledge transfer capability whereby software robots are able to automatically process and transfer knowledge about semi-structured and unstructured data, which is assisted by machine learning and AI-technologies.

> I believe it's over 2000 people of the US federal government who ... figure out what data is being requested, determine whether that data is allowed to be used. So, they have a set of rules that they follow and then take the data, extract the data and then provide the appropriate data back. Our goal is to let robots do that for the majority of those requests. (Respondent R9)

[5]Datarobot.com.

IT knowledge retention capability is the capacity to maintain IT-related knowledge for future use (Li & Chan, 2019). Based on the interview data from respondents R1, R3, and R5, we summarize the impact of RPA on IT knowledge retention capability. Cognitive RPA adds to the IT knowledge retention capability as software robots are able to automatically capture, process, codify, and store data.

5 Discussion

Our aim was to increase the understanding of cognitive RPA and the impacts of RPA on public sector organizations' dynamic IT capabilities. As a result, we contribute with a definition and a conceptualization of the concept of *cognitive RPA* and an analysis of how RPA, including cognitive RPA, affects public sector organizations' dynamic IT capabilities.

5.1 Definition and Conceptualization of Cognitive RPA

We recognize that there are several concepts used to describe the combination of RPA with artificial intelligence, such as hyper automation (Ray et al., 2020) and intelligent RPA (van de Weerd et al., 2021). However, we argue, along the same lines as Viehhauser (2020), that RPA is combined with a particular type of artificial intelligence, often referred to as cognitive intelligence (Davenport & Kirby, 2016). Hence, cognitive RPA would be a natural and more specific term to use.

The concept of cognitive RPA is multidimensional. On a general level, cognitive RPA could be viewed as a category of RPA. On a more specific level, it includes several sub-categories that currently emerge in public organizations: document processing, image processing, conversation processing and process analysis. However, the definition is open for new sub-categories to be added in the future.

The definition aims to extend the view of RPA and to support further studies of the concept. From one perspective, RPA is considered to be augmented by adding artificial intelligence (van de Weerd et al., 2021). From another perspective, held by several of our respondents, it is suggested that RPA provides a vehicle for simplifying the deployment of artificial intelligence in public organizations.

Building on Viehhauser (2020), we contribute with a more comprehensive conceptualization of cognitive RPA, see Fig. 1. Cognitive RPA is conceptualized as an open system that includes interactions with the environment and feedback loops used to evaluate software robots, including training of AI-models.

Also, cognitive RPA currently includes four sub-categories. These four sub-categories depend on the development of AI-technologies carried out by partners to RPA vendors. Therefore, we believe that the cognitive functions will continue to be extended and we expect there to be more sub-categories added to cognitive RPA in the future. As pointed out by Viehhauser (2020), it is debatable how much of the cognitive functions will be included in the RPA vendors' software in the future. However, some respondents believed that more AI capabilities will be integrated soon in the RPA software offered by vendors.

5.2 Propositions for the Impact of RPA on Dynamic IT Capabilities in Public Organizations

Based on the analysis, we propose that RPA affects public organizations' dynamic IT capabilities. RPA affects all three dynamic capabilities and influences the nine ordinary IT-related capabilities in several ways. We make 18 propositions for the relationship between RPA and dynamic IT capabilities, see Table 3. As a category of RPA, we propose that cognitive RPA extends the impact of RPA on dynamic IT capabilities. For example, cognitive RPA increases IT infrastructure capability by enabling automation based on semi-structured and unstructured information. Also, RPA increases IT exploitation capability by providing an environment for exploiting artificial intelligence. Moreover, cognitive RPA affects IT infrastructure flexibility, IT deployment capability and IT exploration capability by discovery, rapid deployment and scaling of automated solutions through process analysis.

The propositions presented in Table 3 can serve as a foundation for further research concerning the relationship between RPA and dynamic IT capabilities in public sector organizations.

5.3 Study Limitations and Future Research

There are several limitations to the study. First, it is based on data provided by vendors of RPA software. On the one hand, they have first-hand knowledge about how their software is used by their customers. On the other hand, they could be biased towards new and less tested versions of their software as well as having a limited understanding of the full socio-technical implications of the use of their software from the perspectives of other stakeholders. The adaptation of cognitive RPA to local circumstances may become an issue that is worthy of investigation. Second, we have uncovered cognitive RPA as a more technical construct. For future research, it would be beneficial to also include the social aspects of the concept. Third, we are limited to an IT-centric view of RPA through the choice of dynamic IT capabilities as lens for studying the impact of RPA on public organizations. For future research it would be valuable to take a socio-technical perspective (Sarker et al., 2019) to better study the general impact of RPA on dynamic capabilities. For example, several respondents perceived significant differences in skills requirements between RPA-developers and data scientists and both skills are required to set up software robots with cognitive intelligence.

Table 3 Propositions for how RPA affects dynamic IT capabilities in public sector organizations

Dynamic IT capability	Ordinary IT capability	Proposition
Dynamic digital platform capability	IT infrastructure capability	1. RPA enables more activities and processes to be automated, including structured and unstructured information 2. RPA vendors offer platforms and cloud services for deploying RPA
	IT integration capability	3. RPA provides means for systems integration
	IT infrastructure flexibility	4. RPA shortens lead times for automation and systems integration 5. RPA increases scalability of automation and integration solutions.
Dynamic IT management capability	IT deployment capability	6. RPA deployment favors federated and agile models guided by centers of excellence 7. RPA supports scalability and versatility through process analysis
	IT exploitation capability	8. RPA adds resources for automation with software robots. 9. RPA simplifies for citizens and civil servants to actively participate in IT exploitation. 10. RPA provides an environment for rapid and simple exploitation of artificial intelligence.
	IT exploration capability	11. RPA makes automation more accessible and affordable for public organizations. 12. RPA supports experimentation and discovery of automated solutions.
Dynamic IT management capability	IT knowledge creation capability	13. RPA can be used to automate knowledge discovery and to enable interactions between humans and machines. 14. Federated RPA deployment models support knowledge creation through feedback loops between local applications and centers of excellence.
	IT knowledge transfer capability	15. RPA vendors provide different resources, such as forums, to share knowledge about RPA applications. 16. Federated RPA deployment models support knowledge sharing about situated RPA applications. 17. RPA can be used to automatically transfer knowledge internally and externally
	IT knowledge retention capability	18. RPA can automatically capture, process, codify, and store data.

Last but not least, we are fully aware that the theoretical understanding of RPA and dynamic IT capabilities that we adapted in this chapter may limit the analysis of the interview results and the conceptualization of cognitive RPA. However, this paper advocates taking an explorative approach to study cognitive RPA and its impacts on public organizations. We shall continue the research from other theoretical aspects, for example, by studying cases of cognitive RPA through the lens of routine capability (Swanson, 2019).

6 Conclusion

We defined cognitive RPA as a category of RPA where software robots are able to capture, process and act on semi- and unstructured information. We used this definition to conceptualize RPA as an open system comprising of input, process, output, and feedback. Furthermore, the definition is open to additional sub-categories, as new combinations of artificial intelligence and RPA emerge in public sector cases as well as when social aspects of cognitive RPA become better understood.

Moreover, we contribute with 18 propositions for how RPA affects dynamic capabilities in public sector organizations. They provide a testable foundation for future research on the relation between RPA and dynamic IT capabilities. In addition to research, the results can also help public sector organizations to be more aware of the differences between RPA, cognitive RPA and artificial intelligence and how these technologies can be combined for different purposes. Also, the propositions may help public sector organizations to better understand how RPA affects their dynamic capabilities and what type of resource that RPA represents.

We conclude by suggesting that further research on the relation between RPA and dynamic IT capabilities in public sector organizations could include evaluation of the propositions presented in Table 3 and investigation of the social aspects of cognitive RPA by taking a more socio-technical perspective on its impacts than we have done.

References

Agostinelli, S., Marrella, A., & Mecella, M. (2019). *Research challenges for intelligent robotic process automation*. Paper presented at the Workshop on Artificial Intelligence for Business Process Management (AI4BPM19) held in conjuction with the 17th Int. Conference on Business Process Management (BPM'19). Vienna, Austria, September 1–6.

Anagnoste, S. (2018). Setting up a robotic process automation center of excellence. *Management Dynamics in the Knowledge Economy, 6*(2), 307–332.

Bharadwaj, A. S. (2000). A resource-based perspective on information technology capability and firm performance: An empirical investigation. *MIS Quarterly, 24*(1), 169–193.

Braun, V., & Clarke, V. (2006). Using thematic analysis in psychology. *Qualitative Research Journal, 3*(2), 77–101.

Collins, J. (2005). *Good to great and the social sectors*. Independent Monograph.

Dahiya, M. (2017). A tool of conversation: Chatbot. *International Journal of Engineering and Computer Science, 5*(5), 158–161.

Davenport, T. H., & Kirby, J. (2016). Just how smart are smart machines? *MIT Sloan Management Review, 57*(3), 21–25.

Denagama Vitharanage, I. M., Bandara, W., Syed, R., & Toman, D. (2020). *An empirically supported conceptualisation of robotic process automation (RPA) benefits*. Paper presented at the 28th European Conference on Information Systems (ECIS2020), Marrakesh, Morocco, June 15–17.

Dias, M., Pan, S., & Tim, Y. (2019). *Knowledge embodiment of human and machine interactions: Robotic-process-automation at the Finland government*. Presented at the 27th European Conference on Information Systems (ECIS2019), Stockholm Uppsala, Sweden, June 8–14.

Dubois, A., & Gadde, L. E. (2002). Systematic combining: An abductive approach to case research. *Journal of Business Research, 55*(7), 553–560.

Eikebrokk, T. R., & Olsen, D. H. (2020). Robotic process automation and consequences for knowledge workers; a mixed-method study. In M. Hattingh, M. Matthee, H. Smuts, I. Pappas, Y. K. Dwivedi, & M. Mäntymäki (Eds.), *Responsible design, implementation and use of information and communication technology* (Vol. 12066, pp. 114–125). Springer International Publishing. https://doi.org/10.1007/978-3-030-44999-5_10

Eisenhardt, K. M., & Martin, J. A. (2000). Dynamic capabilities: What are they? *Strategic Management Journal, 21*(10–11), 1105–1121.

Fereday, J., & Muir-Cochrane, E. (2006). Demonstrating rigor using thematic analysis: A hybrid approach of inductive and deductive coding and theme development. *International Journal of Qualitative Methods, 5*(1), 80–92.

Forrester Research. (2014). Building a Center of Expertise to Support Robotic Automation. A Forrester Consulting Thought Leadership Paper Commissioned by Blue Prism.

Geyer-Klingeberg, J., Nakladal, J., Baldauf, F., & Veit, F. (2018). Process mining and robotic process automation: A perfect match. In *Proceedings of the Dissertation Award, Demonstration, and Industrial Track at BPM 2018*. CEUR Workshop Proceedings (Vol. 2196, pp. 124–131).

Goday-Verdaguer, A., Mannhardt, F., & Torvatn, H. Y. (2020). Mining e-mail conversations to enrich event logs: An exploratory case study of a hiring process in a Norwegian municipality. In *Norsk konferanse for organisasjoners bruk at IT* (Vol. 28, No. 1).

Guest, G., MacQueen, K. M., & Namey, E. E. (2011). Introduction to applied thematic analysis. In V. Knight, A. Virding, A. Rosenstein, & A. Hutchinson (Eds.), *Applied thematic analysis*. Sage Publications, Ltd.

Houy, C., Hamberg, M., & Fettke, P. (2019). *Robotic process automation in public administrations. Digitalisierung von Staat und Verwaltung. Lecture Notes in Informatics* (Vol. 291). Köllen.

Jimenez-Ramirez, A., Reijers, H. A., Barba, I., & Del Valle, C. (2019). A method to improve the early stages of the robotic process automation lifecycle. In P. Giorgini & B. Weber (Eds.), *Advanced information systems engineering* (pp. 446–461).

Kallinikos, J. (2004). Deconstructing information packages – Organizational and behavioural implications of ERP systems. *Information Technology & People, 17*(1), 8–30.

Kaplan, A., & Haenlein, M. (2019). Siri, Siri, in my hand: Who's the fairest in the land? On the interpretations, illustrations, and implications of artificial intelligence. *Business Horizons, 62*(1), 15–25.

Kirchmer, M., & Franz, P. (2019). Value-Driven Robotic Process Automation (RPA). In B. Shishkov (Ed.), *Business Modeling and Software Design (BMSD 2019). Lecture Notes in Business Information Processing* (Vol. 356). Springer. https://doi.org/10.1007/978-3-030-24854-3_3

Lacity, M. C., & Willcocks, L. P. (2016). A new approach to automating services. *MIT Sloan Management Review, 58*(1), 41–49.

Lazarus, S. (2018). *Achieving a successful robotic process automation implementation: A case study of Vodafone and Celonis*. Retrieved June 21, 2021, from https://spendmatters.com/2018/06/07/achieving-a-successful-robotic-process-automation-implementation-a-case-study-of-vodafone-and-celonis/

Li, T. C., & Chan, Y. E. (2019). Dynamic information technology capability: Concept definition and framework development. *The Journal of Strategic Information Systems, 28*(4), 101575.

Lindgren, I. (2020). Exploring the use of robotic process automation in local government. *EGOV-CeDEM-ePart, 2020*, 249.

Martins, P., Sá, F., Morgado, F., & Cunha, C. (2020). *Using machine learning for cognitive Robotic Process Automation (RPA)*. Paper presented at the 15th Iberian Conference on Information Systems and Technologies (CISTI 2020), Sevilla, Spain, 24–27 June.

Mora, M., Gelman, O., Paradice, D., & Cervantes, F. (2008, May). The case for conceptual research in information systems. In *CONF-IRM 2008 Proceedings* (p. 52).

Myers, M. D., & Avison, D. (Eds.). (2002). *Qualitative research in information systems: A reader*. Sage.

Nauwerck, G., & Cajander, A. (2019). *Automatic for the people: Implementing robotic process automation in social work*. Presented at the 17th European Conference on Computer-Supported Cooperative Work (ECSCW 2019), Salzburg, Austria, 8–12 June 2019.

Pablo, A. L., Reay, T., Dewald, J. R., & Casebeer, A. L. (2007). Identifying, enabling and managing dynamic capabilities in the public sector. *Journal of Management Studies, 44*(5), 687–708.

Patil, S., Mane, V., & Patil, P. (2019). Social innovation in education system by using Robotic Process Automation (RPA). *International Journal of Innovative Technology and Exploring Engineering, 8*(11), 3757–3760.

Penttinen, E., Kasslin, H., & Asatiani, A. (2018). *How to choose between robotic process automation and back-end system automation?* Presented at the 28th European Conference on Information Systems (ECIS 2018), Portsmouth, UK, 23–28 June.

Ranerup, A. (2020). *Translating robotic process automation in social work: Aspirational changes and the role of technology*. Presented at the 11th Scandinavian Conference on Information Systems (SCIS 2020) Sundsvall, Sweden, 9–11 August.

Ranerup, A., & Henriksen, H. Z. (2019). Value positions viewed through the lens of automated decision-making: The case of social services. *Government Information Quarterly, 36*(4), 101377.

Ranerup, A., & Henriksen, H. Z. (2020). Digital discretion: Unpacking human and technological agency in automated decision making in Sweden's social services. *Social Science Computer Review*. https://doi.org/10.1177/0894439320980434

Ray, S., Villa, A., Tornbohm, C., Rashid, N., & Alexander, M. (2020). *Magic quadrant for robotic process automation* (Vol. 441474). Gartner Inc.

Sarker, S., Chatterjee, S., Xiao, X., & Elbanna, A. (2019). The sociotechnical axis of cohesion for the IS discipline: Its historical legacy and its continued relevance. *MIS Quarterly, 43*(3), 695–720.

Schultze, U., & Avital, M. (2011). Designing interviews to generate rich data for information systems research. *Information and Organization, 21*(1), 1–16.

Singh, A., Ramasubramanian, K., & Shivam, S. (2019). *Building an enterprise chatbot: Work with protected enterprise data using open source frameworks*. Apress Media LLC.

Stolpe, A., Steinsund, H., Iden, J., & Bygstad, B. (2017, November). *Lightweight IT and the IT function: Experiences from robotic process automation in a Norwegian bank*. Presented at the Norsk konferanse for organisasjoners bruk at IT (NOKOBIT 2017), Oslo, Norway, 27–29 November.

Suri, V. K., Elia, M. D., Arora, P., & van Hillegersberg, J. (2018). *Automation of knowledge-based shared services and centers of expertise*. Presented at the 12th Global Sourcing Workshop, La Thuile, Italy, 21–24 February.

Swanson, E. B. (2019). Technology as routine capability. *MIS Quarterly, 43*(3), 1007–1024.

Syed, R., Suriadi, S., Adams, M., Bandara, W., Leemans, S. J., Ouyang, C., ... Reijers, H. A. (2020). Robotic process automation: Contemporary themes and challenges. *Computers in Industry, 115*, 103162.

Teece, D. J., Pisano, G., & Shuen, A. (1997). Dynamic capabilities and strategic management. *Strategic Management Journal, 18*(7), 509–533.

van de Weerd, I., Nieuwenhuijs, B., Bex, F., & Beerepoot, I. (2021). *Using AI to augment RPA: A conceptual framework*. Presented at the 29th European Conference on Information Systems (ECIS 2021), Marakesh, Morocco, 14–16 June.

Van Der Aalst, W. (2012). Process mining. *Communications of the ACM, 55*(8), 76–83.

van der Aalst, W. M. (2020). On the Pareto principle in process mining, task mining, and robotic process automation. In *DATA* (pp. 5–12).

van der Aalst, W. M., Bichler, M., & Heinzl, A. (2018). Robotic process automation. *Business & Information Systems Engineering, 60*(4), 269–272.

Viehhauser, J. (2020). *Is robotic process automation becoming intelligent? Early evidence of influences of artificial intelligence on robotic process automation*. Presented at the 18th International Conference on Business Process Management (BPM 2020), Sevilla, Spain, 13–18 September.

von Bertalanffy, L. (1968). *General system theory: Foundations, development, applications*. George Braziller.

Willcocks, L., Lacity, M., & Craig, A. (2015). *The IT function and robotic process automation*. The Outsourcing Unit Working Research Paper Series, London School of Economics and Political Science, UK.

Part III
Applications of Public Sector Service Automation

Automation and Public Service Values in Human Resource Management

Marcus Persson and Andreas Wallo

1 Introduction

Robot Process Automation (RPA) refers to a broad class of technologies that allow a computer to perform tasks that typically require human cognition, including adaptive decision-making. Automated IT systems are used with the intent to streamline work tasks to reduce cost, improve quality, and speed up repetitive and rule-driven processes in organizations (Anagnoste, 2017). Because of the repetitive and resource-demanding nature of work tasks, the Human Resource (HR) function has been pinpointed as a suitable candidate for automation (Baakeel, 2020; Tambe et al., 2019). The HR function is responsible for operative and strategic work with human resource management (HRM), which involves strategic and operational work with the employment, development, and well-being of the people working in organizations (Armstrong, 2017).

The idea that HRM processes and work tasks could become digitalized is not new and the concept of electronic HRM (e-HRM) has been in use since the late 1990s. Since that time, several somewhat loosely related concepts have gathered under the umbrella term digital HRM, the most recent of which is algorithmic (Cheng & Hackett, 2021; Meijerink et al., 2021).

HRM work tasks that have been identified as potentially suitable for various degrees of automation include staffing, recruitment and onboarding, performance management, payroll and worktime management, and administration (Anagnoste, 2017; Balasundaram & Venkatagiri, 2020; Upadhyay & Khandelwal, 2018). For example, there are programs that not only help to create and publish job descriptions for applications, but also use bias-free, gender-neutral language that can be directed to a concrete target group. Another example is a digital applicant tracking system

M. Persson (✉) · A. Wallo
Department of Education and Sociology, Linköping University, Linköping, Sweden
e-mail: marcus.persson@liu.se; andreas.wallo@liu.se

G. Juell-Skielse et al. (eds.), *Service Automation in the Public Sector*, Progress in IS,
https://doi.org/10.1007/978-3-030-92644-1_5

that can pre-screen applications, identify keywords and match candidates with the right positions. Additionally, instead of having recruiters conducting several repetitive phone/video screenings per day, a company can introduce a chatbot that will replace a human interviewer. Once recruited, activities connected to the onboarding of a new employee can also be automated, for instance, through tutorials regarding various systems. By automating work processes, time and resources can be freed up so that HR professionals can focus on more advanced and strategically oriented work tasks.

Several reviews of the literature on the digitalization of HRM have been published during the last decade, and these have focused on (for example) the transformational potential of information and communications technology for organizational development and revenue optimizing (Wirtky et al., 2016; Zeebaree et al., 2019), effectivization and the speeding up of processes as well as improving quality of work (Strohmeier, 2009), organizational value creation and potential competitive advantages (Zeebaree et al., 2019), strategic HRM and strategic outcomes at the macro-level (Marler & Fisher, 2013), and what the utilization of intelligent automation means for HRM (Vrontis et al., 2021). However, the results from these previous reviews are mixed and depend heavily on factors relating to the specific national and organizational context, such as the type of organizations or type of ICT solutions used.

This chapter will provide an updated overview of the field, including the latest technological solutions relevant for the transformation of HRM. Moreover, the chapter will also propose an organizational contextualization in the form of RPA usage within public service organizations. It is important to account for the specific organizational rules and values governing public sector organizations because, in comparison with private organizations, public sector organizations are governed by values and regulations that aim to promote the public good, democratic values, and be of service of all members of society (not only those paying for the services provided by a private organization) (Wirtz et al., 2019). It is important to recognize that public sector organizations are governed by specific values intended to uphold public trust (Ranerup & Henriksen, 2019; Sundberg, 2019), which is why automation in public sector organization cannot simply be reduced to a matter of enhancing efficiency according to economic value. Consulting previous reviews, we found only one contribution (Mozgovoy & Mettler, 2019) with an explicit interest in e-HRM within public sector organizations.

This chapter will explore empirically grounded issues that are evoked by utilizing RPA in public sector organizations and will emphasize the importance of public service values, thereby offering a nuanced understanding of the implication of RPA for HRM in public sector organizations. More specifically, the findings of the reviewed literature will be accounted for according to public service values, and potential changes to the professional role and competencies of HR practitioners.

We will begin by describing our theoretical points of departure. Next, we will outline the methodology used to carry out the literature review. We will then present the findings of our review and discuss them in the light of the theoretical framework.

Finally, we will present our conclusions and suggest implications for future research and practice.

2 Automation and Public Service Values

In order to better comprehend the implications for HRM in the specific context of public sector organizations, we will draw upon the theory of public service values (DeForest Molina & McKeown, 2012; Demir et al., 2015). By adopting the theoretical notion of public service values, we intend to create a better understanding of how automating technology could affect the collective values that are at the heart of public service organizations.

The theoretical interest in public service values has grown since the early 1980s and can be traced back to the "corporate culture" movement, which emphasized the connections among an organization's success, its values, and its corporate culture (see, e.g., Schein, 2017). A second influential movement was New Public Management (NPM), which called for applications of business practices and technologies in the public sector (Hood, 1991). NPM affected public service values by introducing strategic plans with a mission, a vision, a values statement, an increased focus on accountability for results, and an emphasis on business values such as innovation and service (Kernaghan, 2003).

These two movements have contributed to the proliferation of ethics and values documents that are characteristic of today's public sector organizations. Although it is possible to identify a wider range of organizational values, our conceptualization builds on Kernaghan's (2003) discussion of ethical, democratic, professional, and people values. According to Kernaghan, this categorization should not be seen as restrictive or exclusive; instead, he argues that some values fall into more than one category. In addition, there is also an association between ethical and democratic values. In this review, we adapt Kernaghan's model to fit the empirical area of HRM, and mainly focus on three categories of values as presented in Fig. 1.

Because the three categories are not mutually exclusive, value conflicts within and between the categories are inevitable. As a result, public administrators in a given situation are often faced with multiple possible actions, each justified by different public service values (Turner, 2015). Values help explain why managers act or choose not to act in certain ways, their specific decisions, and their understanding of administrative responsibilities within the broader scheme of government (Rokeach, 1973; Witesman & Walters, 2013).

Several researchers have highlighted the relationship between discretionary HR practices and job crafting among public sector employees (Luu, 2020). It has been suggested that HR practices may help public sector employees to identify with the goals (i.e., public service values) of their organization, enhance their competencies and empower them, and create conditions that influence public administrators to proactively engage in crafting their public service job (Gavino et al., 2012).

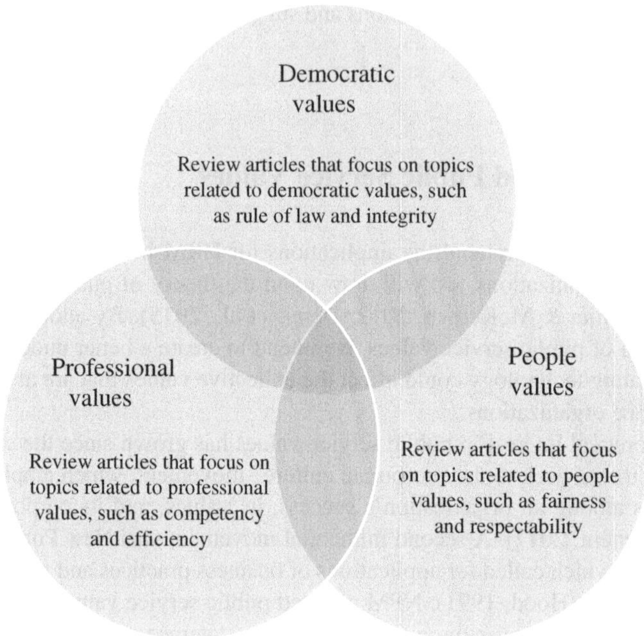

Fig. 1 Three categories of public service values (based on Kernaghan, 2003)

Against this background, we ask what changes RPA brings for HRM concerning the professional role and competencies of HR practitioners. It is important to remember that modern HR is a relatively new occupation that some argue is struggling with issues of legitimacy (Alvesson & Lundholm, 2014; Kochan, 2007). The ability of HR practitioners to deliver services that add value to their customers, typically a line manager or middle manager, and act as business partners has been debated (Heizmann & Fox, 2019). On this note, there is an ongoing discussion in the HR literature regarding the professional competencies needed when work routines and work tasks change due to automation and technological advancements, and how this development may affect the status and legitimacy of HR professionals in organizations (Cooke et al., 2020; Ulrich et al., 2013). Prominent topics in the current debate are whether the automated processes will replace or complement the competencies of the HR practitioners (Rana, 2018) and whether the automation of simple operational HR tasks will increase the time available for strategic HR issues (Bondarouk & Ruël, 2013; Obeidat, 2016).

3 Method and Data

The method is based on a scoping literature review (Grant & Booth, 2009). While systematic reviews usually have a quite specific research question, the research questions in scoping reviews are usually broader. Scoping reviews also include existing literature and findings from a range of different study designs and methods, which makes the use of formal meta-analytic methods difficult (Sucharew & Macaluso, 2019). We draw on a scoping review method to highlight a new and under-researched area by critically assessing studies and identifying new insights and knowledge gaps (Munn et al., 2018). The review adhered to the PRISMA guidelines provided by Moher (2009).

The following inclusion criteria were used to select articles for review: (a) peer-reviewed, (b) written in English, (c) reporting findings from empirical studies, and (d) a focus on RPA and HRM. Review articles, pilot studies, conference papers, and papers from sources of unknown quality (like summaries or keynotes) were excluded. Furthermore, although we recognize the existence of many articles that deal with automation in private sector organizations, these articles were excluded due to the review's specific aim to examine the use of RPA in public sector organizations. Descriptive technological or computational papers without any empirical testing have also been excluded.

The literature search was conducted in March 2021 using the database Scopus. The keywords used were combinations of HRM terms ("human resource management," "recruitment," "employee selection," "HR analytics," "HR services") with terms aimed at capturing automation of HR tasks and processes ("robotic process automation," "artificial intelligence," "business process automation," "digital workforce," "digital workers," "workforce automation," "digital HRM," "recruitment automation," "e-HRM," and "AI recruitment").

The initial searches identified 381 papers, and during these initial searches, 3 additional papers were found that we deemed relevant (but that the search had not returned), for a total of 384 records evaluated. We subsequently scanned these for relevance based on keywords, abstracts, and titles, resulting in 55 papers. The authors read the full papers independently to reduce subjectivity. Differences in decisions were discussed and resolved. The final step of the evaluation process was looking at key results, limitations, and interpretation of the results, and after this final evaluation, 22 studies remained (Table 1 in the appendix). The final collection of papers was analyzed through narrative synthesis, which involves central information about the studies being compiled in text and tables (Grant & Booth, 2009).

In the following sections, the findings from the final collection of papers will be presented according to the values discussed above, namely democratic, people, and professional values (Fig. 2).

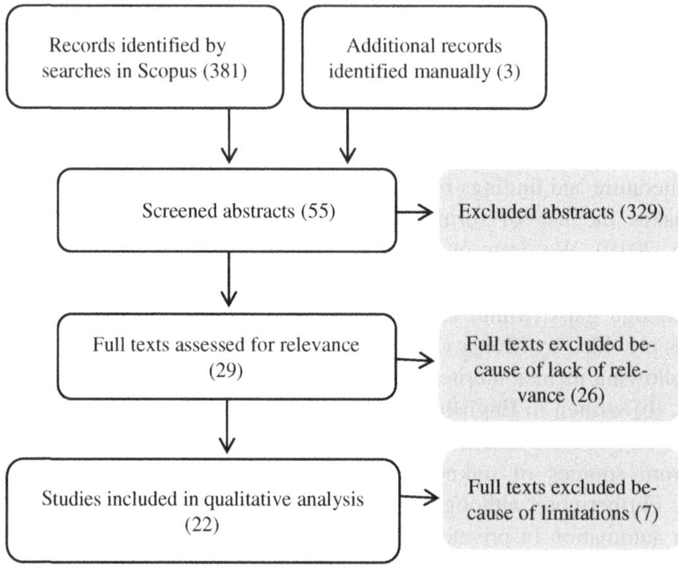

Fig. 2 Prisma flowchart for the systematic literature review process

4 Democratic Values: Commitment to Rules of Law and Integrity of Citizens

In line with the model presented in Fig. 1, we argue that democratic values are related to the ideals of following the rules of law objectively and neutrally while also maintaining integrity. In this regard, our findings indicated possible benefits, barriers, and risks related to the implementation of automated IT systems in HRM. For example, in one recent study (Rahman et al., 2018), some interviewees in one study expressed concerns about the security system and the sensitive data of citizens, which relates to democratic values. Public sector organizations often store and handle sensitive and classified data of citizens and the risk of hacker attacks or data information leaks may have serious consequences for those citizens. Data security is a prioritized issue for organizations in the public sector, even more so when the organizations implement automatized software systems. When IT protocols start to communicate with each other autonomously—for example, by accessing data from different databases and creating new information flows—the issue of data security becomes even bigger. Rahman et al. (2018) stated that data security and risks of information leaks may affect the employees' trust in the system.

Similar considerations were expressed by Oberst et al. (2020), which compared the preferences among HR professionals involved in the recruitment process. These findings indicated that HR recruiters preferred candidates to be recommended by co-workers rather than by a hiring algorithm. In other words, HR professionals trusted their colleagues more than the automated recruitment process. This is paradoxical, because the algorithms only act on old data and established norms, and as such are not more biased than human recruiters. The purpose of using recruiting algorithms is to neutralize such biases, but if the algorithms are also perceived to be biased, humans seem to prefer to trust a biased human rather than a biased machine.

The reviewed literature highlighted the significance of data security and the relationship between integrity and trust. Ultimately, democratic values in public sector organizations seem to be deeply connected to the citizens' trust in the state system.

Related to this type of values are also HR practices concerning issues of commitment to the organizational culture of the public sector, i.e., commitment to public service values (Nura & Osman, 2013; Reddy et al., 2019). An example is Roy and Jegan's (2019) comparative study of organizational values and commitment among banks in the public and private sectors. This study revealed that banks in the public and private sectors differed concerning their background and organizational commitment, and that the degree of implementation of e-HRM practices was higher in private sector banks than public sector banks. The level of organizational commitment in private sector banks was also higher than that of public sector banks. Furthermore, e-recruitment, e-training, and e-information sharing were all found to significantly influence the organizational commitment at banks.

Taken together, these findings may suggest that there is a connection between the use of automated e-HRM services and the commitment of the employee: e-recruitment may be used to target the right candidates for the organization; e-training and e-information sharing may be used to educate through internal training and cultivating organizational values. Although various kinds of automated services may function well to promote organizational values, it should be noted that there may be a risk of diminishing certain values if they are already firmly established in the organization. For example, Rahman et al. (2018) found that the use of RPA can improve communication between HR professionals and employees, but they also expressed concerns about the risk of losing social contact between HR professionals and employees, which may negatively affect job satisfaction and organizational commitment.

5 People Values: Fairness and Bias in Recruitment

In the reviewed studies, the people values (Kernaghan, 2003) were specially addressed in relation to avoiding bias in recruitment and selection processes (Furtmueller et al., 2011; Hooper et al., 1998; Köchling et al., 2021; Oberst et al., 2020). In the HRM literature, it is assumed that when recruiters follow their intuition, they may unwittingly consider factors that are irrelevant to future job performance, such as age, sex, or race of the applicants (Highhouse, 2008). By using algorithms, the recruitment process can focus on matching competency with organizational needs in an instrumental manner. For example, Köchling et al. (2021) studied the use of algorithmic video analysis in the recruiting context. The applicants recorded a video of themselves, which was then evaluated. This method had both practical and economic benefits because recruiters became more efficient in handling and screening applicants in less time, which, in turn, reduced the time-to-hire.

Moreover, organizations want to increase the objectivity and fairness of the recruitment process by implementing algorithmic decision-making and seeking to diminish human bias (e.g., prejudices and personal beliefs). Fairness in recruitment should be understood at both the individual and group levels. Individual fairness ensures that any two individuals with equal competencies and formal merits should be classified similarly. Group fairness (that is, statistical parity) ensures that overall positive (and negative) classifications are similar for specific groups and the overall population. By introducing RPA, the hope is to increase fairness in the recruitment process, thereby recruiting the most competent employees and increasing the diversity in the workplace.

However, using RPA for recruitment also presents some problems. Machine learning systems are often left to act on old data and social norms that govern the organization (Tambe et al., 2019). For example, the study by Köchling et al. (2021) showed that the under-representation of genders and various ethnicities in the training data set leads to an unpredictable overestimation and/or underestimation of the likelihood of inviting representatives of these groups to a job interview. In other words, algorithms were replicating the existing inequalities in the data set. Overall, several of the articles (e.g., Furtmueller et al., 2011; Oberst et al., 2020) suggested that the full potential of online recruiting was not being realized due to ongoing difficulties in unbiased matching and searching of candidates.

The literature review revealed that RPA was not only used to recruit new employees but also to assess and select existing employees for job promotion and other positions of career development in organizations (Hooper et al., 1998; Reddick, 2009; Tambe et al., 2019). Online selection systems have, in this regard, become important tools to increase the likelihood of incumbents meeting role requirements, and whether those requirements are met is assessed through various tests. For example, Hooper et al. (1998) studied the use of a basic rule-based expert IT system in the selection of candidates for an army officer's educational program. The results indicated that the computer system decisions were not statistically different from the decisions made by human experts. This finding may be interpreted in line with the above logic that recruitment and selection algorithms are based on old data and prevailing organizational norms.

6 Professional Values: Efficient Service and/or Innovative Support

One of the most convincing arguments for using RPA in public sector organizations is to enhance efficiency. But what does this mean for the service provided by the HR function? Much of the existing research about the use of automating technologies addressed issues related to efficiency and service quality (Ibrahim et al., 2018;

Nurlina et al., 2020; Obeidat, 2016; Reddick, 2009; Wahyudi & Park, 2014). As such, the main body of research within the field is related to professional values.

Several studies have explored the relationship between automating technologies and efficiency. The results are mixed depending on what processes the organization were trying to make more efficient. For instance, from the HR specialist point of view, online training is now regarded as an efficient way to dispense training within public sector organizations, reducing direct costs for instructors, printed materials, and training facilities, and also indirect costs related to travel time, lodging and travel expenses (Tambe et al., 2019). The services targeted for automation in the reviewed studies were often administrative and the intent with automation was mainly to improve efficiency. For example, Obeidat (2016) studied HR activities, such as payroll, performance appraisal, benefits administration, HRM information system, learning management, self-service, and absence management, and found that e-HRM positively contributed to HRM effectiveness. Similarly, Nura and Osman (2013) found a connection between the application of HR technology and employee retention, which led to fewer complaints, reduced number of conflicts, and increased job satisfaction. It is possible to interpret these and other studies (Wahyudi & Park, 2014; Nurlina et al., 2020) as examples of administrative services that are suitable for automation, and where self-service might even be preferable to services provided by human HR professionals.

As stated above, it is important to consider the nature of the provided services by the HR function to give a nuanced understanding of the relationship between efficiency and the type of provided service. A work by Reddick (2009) studied the implementation of Web-based self-services in city governments. The results revealed that the Web was merely being used as an information source for managers and employees, and there was limited evidence of more advanced self-service applications. In a more recent study by Ibrahim et al. (2018), the structural relationships between organizational information system-related support and end-user satisfaction with e-HRM in government agencies were investigated. The results showed that literacy support and technical support had significant and positive relationships with end-user satisfaction with the system. Innovation support did not show any significant contribution. Not surprisingly, technical support was fundamental to user satisfaction: if the system did not function as planned, the user would react to it negatively.

More interesting was the correlation between user satisfaction and literacy support (which here refers to mechanisms that educate through sharing of system-related information, like training and documentation). The only aspect that did not show any significant contribution was innovation support, which is the mechanism encouraging professionals to experiment and learn. This mechanism required management to take several actions so that users could be more innovative and explorative in using the system, such as promoting supportive relationships among employees, facilitating communication and discussion, encouraging new ideas, and providing incentives for learning (Baldegger et al., 2020; Ibrahim et al., 2018).

In sum, automating IT systems can be effective for creating efficiency in administrative and information-based services provided by HRM. But when it comes to more innovation-oriented services, we find little support for the benefits of RPA. Most likely this has to do with the fundamental sociopsychological requirements of creativity at the heart of innovative thinking and working. Creativity is seldom linear but rather a dynamic (back-and-forth) process, in which the thoughts of two or more actors feed into each other and together create something that could not have been conceived without the cooperation of the others. Creative dynamics is difficult to automate in any linear processes, and still requires a competent human other.

7 Toward a More Strategic Professional Role for HR?

RPA is expected to enhance the operational capabilities of the HR function by simplifying administrative processes, reducing the time required for HR transactions, tracking job records, and managing the employee payroll and benefits programs. Some scholars argue that automating IT systems will offer opportunities for the HR function to reduce the time spent on administrative work, and instead focus on developing practices, designing HR policies, business planning, performance, succession planning, and building human capital (Marler & Parry, 2016). In other words, digital automation is expected to act as an enabler of the dynamic capabilities of the HR function.

However, when reviewing the literature, we found little support for the idea that the role of is HRM changing. Despite the great potential for HR functions to create value beyond administrative outcomes, in practice, many organizations seem to utilize e-HRM more for an automating approach that focuses primarily on administrative efficiency rather than supporting strategic human capital management processes (Baldegger et al., 2020; Rahman et al., 2018; Strohmeier, 2020; Troshani et al., 2011).

A well-cited study in this regard was conducted by Bondarouk and Ruël (2013), which found that e-HRM did not automatically result in any direct strategic benefits. Only if certain conditions were met could the dynamic and operational capabilities of the HR function be strengthened. The researchers stated that while HR professionals perceived role changes as the result of e-HRM, line managers and non-managerial employees did not. This result was confirmed in a study by Wayudi and Park (2014), in which the researchers could not find any significant relationship between e-HRM usage and a strategic role shift of HRM function. Instead, they found that implementation of the e-HRM system improved HR's administrative efficiency and achieved cost reductions. Wayudi and Park stated that if e-HRM is introduced with the main purpose of establishing additional infrastructures but lacks a strong alignment with the business, it is unlikely that the strategic value of e-HRM will be realized.

The implementation of technology, by itself, does not appear to transform and add strategic value to the role of HRM (Baldegger et al., 2020). Instead, technology serves as a tool of automation and cost reduction within the administrative areas (Rahman et al., 2018). If RPA is simply used to automate administrative work tasks performed by HR professionals, it does not alter the role of HR in any significant way. Automation can only make processes and work routines more efficient; it cannot create new processes and new work routines. This does not mean that implementing RPA could not enable a new, more strategic role for HRM. But for this to happen, the organization needs a set of internal and external resources as well as traditional HRM systems that could initiate, implement, and actualize organizational strategies per se (Marler & Parry, 2016).

It has also been argued that new ways of working will lead to the need for a new set of professional skills, especially if the role of HR changes into a more strategic, business-oriented role. Working more strategically would probably demand more social, economic, entrepreneurial, and managerial competencies. Some scholars have suggested that the implementation of RPA will result in an "upskilling" of employees' professional skills (Johansson et al., 2020; Vallor, 2015), and it could be argued that successful implementation of automated IT systems will require a new set of skills (Strohmeier, 2020; Troshani et al., 2011). In other words, the argument that implementation of RPA will lead to an upskilling may prove legitimate. However, this upskilling may not be characterized by more advanced skills associated with a more strategic role of HRM, but rather with a different set of skills needed to work with an automated IT system. In this regard, it would probably be more accurate to talk about re-skilling rather than up-skilling.

8 Discussion

In this chapter, we have identified and discussed empirically grounded issues that are actualized when RPA is utilized in public sector organizations. The main findings of the review are that democratic values are related to the importance of upholding and safeguarding data security and the integrity of citizens. IT systems must enable HR professionals to work according to their commitment to the rules of law. Ultimately, democratic values in public sector organizations seem to be deeply connected to citizens' trust in the state system.

Concerning trust, organizations may want to increase the objectivity and fairness of the recruitment process by implementing algorithmic decision-making and seeking to diminish human bias (e.g., prejudices and personal beliefs). Findings show that algorithms often replicate existing inequalities in a data set and continue to reproduce biased matching and searching of candidates.

The main body of research within the field related to professional values and addressed issues related to efficiency and service quality. Findings indicated that automating IT systems can be an effective way to creating efficiency in administrative and information-based services provided by HRM. But when it comes to more innovation-oriented services, RPA seems to have few if any benefits.

Some scholars argue that creating efficiency in administrative processes will offer opportunities for the HR function to spend less time on administrative work and instead focus on more dynamic capabilities. However, we find little support for the idea that the role of HRM is changing. Many organizations seem to utilize e-HRM more for an automating approach that focuses primarily on increasing administrative efficiency rather than supporting strategic human capital management processes.

New ways of working will likely lead to the need for a new set of professional skills. However, new ways of working may not be characterized by more advanced professional skills but rather with the skills necessary to work with automated IT systems.

Against the backdrop of the theory of public sector values (DeForest Molina & McKeown, 2012; Demir et al., 2015; Kernaghan, 2003), we conclude that previous studies account for HR tasks associated with democratic, people, and professional values. However, Kernaghan's ethics category was much harder to discern in the reviewed articles. The importance of ethical values is often under-served in other research areas that also deal with the implementation of new technological IT solutions. A similar lack of attention to ethical values can, for example, be observed in research regarding the implementation of technology in care work (Persson et al., 2021). The reason for this lack of attention to ethics is possibly because the organizational driver for implementing new technological systems is to create efficiency and reduce time and costs.

Having said this, ethical values are important to uphold in the practical work of civil servants and should not be downplayed. When working in public sector, one is working for the public and not for a company (EU, 2017). Private sector organizations obviously must also follow the rule of law, but their existence depends on economic profit in a way that differs from public sector organizations. Conversely, public sector organizations are more dependent on the citizens' trust in the state system than private sector organizations are (Parker & Bradley, 2000).

The omission of the ethics category points to a need for future research aimed at capturing ethical dilemmas that may arise following the application of RPA in public sector organizations. Other areas where more research would be warranted include an in-depth study of how RPA in HRM affects the competence requirements of HR professionals. The findings of our review only skim the surface in this regard, and further investigations built on a competent theoretical perspective could cast additional light on the question of the re-skilling of HR professionals. We also see a need for further comparative studies regarding the implementation of RPA in both public and private sector organizations in order to identify sector-specific aspects and consequences.

We hope that the scoping review (Grant & Booth, 2009) used as a basis for this chapter has helped highlight the need for further studies of RPA usage in public sector HRM. The total number of analyzed texts indicates that the field is still in an early stage, however, we predict that the number of publications reporting empirical studies will grow significantly in the coming years

Appendix

Table 1 Overview of the reviewed articles

Author(s), Year, Country	Aim of the study	Method(s) and data	Public sector values	Roles and competences
Baldegger et al. (2020) Switzerland	To study the adoption of RPA in HRM processes	Survey involving 310 HR professionals	Professional values: Efficiency and services	Role: Operative and strategic
Bondarouk and Ruël (2013) Netherlands	To study the strategic value of e-HRM	Survey involving 219 employees + 13 interviews	Professional values: Efficiency and services	Roles: Operative and strategic
Bondarouk et al. (2017) Belgium	To study the impact of e-HRM	Case study involving 140 respondents	Professional values: Efficiency and services	
Dilu et al. (2017) Ethiopia	To study implementation of e-HRM	Survey (246 HR employees); qualitative interviews (16)	Professional values: Efficiency and services	Skills and competences
Furtmueller et al. (2011) Netherlands	To explore off- and online recruitment	Interviews with recruiters; analysis of 40 recruiting sites	People values: Fairness and bias in recruitment	
Hooper et al. (1998) USA	To test a system to screen personnel records	Personnel data correlated to votes by 10 human experts	People values: Fairness and bias in selection	
Ibrahim et al. (2018) Malaysia	To study end-user satisfaction	Survey involving 490 employees	Professional values: Efficiency and services	Skills and competences
Köchling et al. (2021) Germany	To study algorithmic decision making	Analysis of a data set of 10,000 video clips of self-presentations	People values: Fairness and bias in recruitment	
Marler and Parry (2016) International	To examine e-HRM and the role of HR	Data set analysis (5665 companies from 32 countries)		Roles: Operative and strategic
Nura and Osman (2013) Nigeria	To study performance, e-HRM, and retention	Survey (196 respondents in higher education)	Democratic value: Commitment Professional values: Efficiency and services	
Nurlina et al. (2020) Indonesia	To analyze e-HRM and employee performance	Survey involving 200 civil servants	Professional values: Efficiency and services	

(continued)

Table 1 (continued)

Author(s), Year, Country	Aim of the study	Method(s) and data	Public sector values	Roles and competences
Obeidat (2016) Jordan	To study e-HRM and its effectiveness	Survey involving 121 employees	Professional values: Efficiency and services	
Oberst et al. (2020) Spain	To examine decision-making in recruitment	Survey involving 135 HR recruiters	Democratic values: Trust People values: Fairness and bias in recruitment	
Rahman et al. (2018) Bangladesh	To investigate the implementation of e-HRM	Interviews with 30 employees	Democratic values: IT security and integrity of citizens Professional values: Efficiency and services	Roles: Operative and strategic
Reddick (2009) USA	To examine the use of e-HRM in public service organizations	Survey among 88 HR professionals	People values: Fairness and bias in recruitment and selection Professional values: Efficiency and services	Competences and roles
Reddy et al. (2019) India	To explore implications of HR analytics and AI for HRM	Analysis (1470 observations) of a structured data set	Democratic value: Commitment, job satisfaction, retention	
Roy and Jegan (2019) India	To study the use of e-HRM and commitment in public and private banks	Interviews with 215 employees	Democratic value: Commitment, communication, job satisfaction People values: Fairness and bias in recruitment	
Strohmeier (2020) Germany	To explore IoT in HRM	Delphy study with 40 IoT-experts	Professional values: Efficiency and services	Skills and competencies
Tambe et al. (2019) USA	To explore challenges in using AI for HR tasks	A short survey, and interviews with experts from 20 major organizations	People values: Fairness and bias in recruitment and selection Professional values: Efficiency and services	

(continued)

Table 1 (continued)

Author(s), Year, Country	Aim of the study	Method(s) and data	Public sector values	Roles and competences
Tansley et al. (2014) UK	To examine efficiency and innovation in e-HRM	A case study involving 12 employees		Roles: Operative and strategical
Troshani et al. (2011)	To study the adoption of RPA	Interviews with 16 employees	Professional values: Efficiency and services	Skills and competencies
Wahyudi and Park (2014) Indonesia	To study value creation in e-HRM	Quantitative survey involving 306 civil servants	Professional values: efficiency and service quality	Roles: Operative and strategic

References

Alvesson, M., & Lundholm, S. (2014). *Personalchefers arbete och identitet: Strategi och strul.* Studentlitteratur.

Anagnoste, S. (2017). Robotic Automation Process—The next major revolution in terms of back office operations improvement. *Proceedings of the International Conference on Business Excellence, 11*(1), 676–686. https://doi.org/10.1515/picbe-2017-0072

Armstrong, M. (2017). *Armstrong's handbook of human resource management practice* (14th ed.). Kogan Page Limited.

Baakeel, O. (2020). The association between effectiveness and Human Resource Management functions and the use of Artificial Intelligence. *International Journal of Advanced Trends of Computer Science and Engineering, 9*(1), 606–612.

Balasundaram, S., & Venkatagiri, S. (2020). A structured approach to implementing Robotic Process Automation in HR. *Journal of Physics: Conference Series, 1427*, 012008. https://doi.org/10.1088/1742-6596/1427/1/012008

Baldegger, R., Caon, M., & Sadiku, K. (2020). Correlation between entrepreneurial orientation and implementation of AI in human resources management (HRM). *Technology Innovation Management Review, 10*(4), 72–79.

Bondarouk, T., & Ruël, H. (2013). The strategic value of e-HRM: Results from an exploratory study in a governmental organization. *The International Journal of Human Resource Management, 24*(2), 391–414. https://doi.org/10.1080/09585192.2012.675142

Bondarouk, T., Harms, R., & Lepak, D. (2017). Does e-HRM lead to better HRM service? *The International Journal of Human Resource Management, 28*(9), 1332–1362.

Cheng, M. M., & Hackett, R. D. (2021). A critical review of algorithms in HRM: Definition, theory, and practice. *Human Resource Management Review, 31*(1), 100698. https://doi.org/10.1016/j.hrmr.2019.100698

Cooke, F. L., Dickmann, M., & Parry, E. (2020). Important issues in Human Resource Management: Introduction to the 2020 review issue. *The International Journal of Human Resource Management, 31*(1), 1–5.

DeForest Molina, A., & McKeown, C. (2012). The heart of the profession: Understanding public service values. *Journal of Public Affairs Education, 18*(2), 375–396.

Demir, T., Reddick, C. G., & Nank, R. (2015). The relationship between public service values and administrative involvement in policymaking. *Public Organization Review, 15*(1), 79–98.

Dilu, E., Gebreslassie, M., & Kebede, M. (2017). Human Resource Information System implementation readiness in the Ethiopian health sector: a cross-sectional study. *Human resources for health, 15*(1), 1–10.

European Union. (2017). *Quality of public administration. A toolbox for practitioners*. Publications Office of the European Union.

Furtmueller, E., Wilderom, C., & Tate, M. (2011). Managing recruitment and selection in the digital age: e-HRM and resumes. *Human Systems Management, 30*(4), 243–259.

Gavino, M. C., Wayne, S. J., & Erdogan, B. (2012). Discretionary and transactional human resource practices and employee outcomes: The role of perceived organizational support. *Human Resource Management, 51*(5), 665–686.

Grant, M. J., & Booth, A. (2009). A typology of reviews: An analysis of 14 review types and associated methodologies: A typology of reviews. *Health Information & Libraries Journal, 26*(2), 91–108. https://doi.org/10.1111/j.1471-1842.2009.00848.x

Heizmann, H., & Fox, S. (2019). O Partner, Where Art Thou? A critical discursive analysis of HR managers' struggle for legitimacy. *The International Journal of Human Resource Management, 30*(13), 2026–2048. https://doi.org/10.1080/09585192.2017.1314974

Highhouse, S. (2008). Stubborn reliance on intuition and subjectivity in employee selection. *Industrial and Organizational Psychology, 1*(3), 333–342. https://doi.org/10.1111/j.1754-9434.2008.00058.x

Hood, C. (1991). A public management for all seasons? *Public Administration, 69*(1), 3–19. https://doi.org/10.1111/j.1467-9299.1991.tb00779.x

Hooper, R. S., Galvin, T. P., Kilmer, R. A., & Liebowitz, J. (1998). Use of an expert system in a personnel selection process. *Expert systems with Applications, 14*(4), 425–432.

Ibrahim, H., Shamsuddin, F. M., Yusoff, Y. M., & Zin, M. L. M. (2018). The structural relationship between organisational-information system related support, technology self-efficacy, and end-user satisfaction with E-HRM. *The Journal of Social Sciences Research*, 1111–1119.

Johansson, B., Engberg, S., Lindgren, I., & Sordal, T. (2020). Impacts of RPA on employees' skill variety. In *AMCIS 2020 Proceedings., Electronic Network, August 10–14*.

Kernaghan, K. (2003). Integrating values into public service: The values statement as centerpiece. *Public Administration Review, 63*(6), 711–719.

Kochan, T. (2007). Social legitimacy of the human resource management profession: A U.-S. perspective. In P. Boxall, J. Purcell, & P. Wright (Eds.), *The Oxford handbook of human resource management*. Oxford University Press.

Köchling, A., Riazy, S., Wehner, M. C., & Simbeck, K. (2021). Highly accurate, but still discriminatory. *Business & Information Systems Engineering, 63*(1), 39–54.

Luu, T. T. (2020). Activating job crafting in public services: The roles of discretionary human resource practices and employee use of normative public values. *Public Management Review*, 1–33.

Marler, J. H., & Fisher, S. L. (2013). An evidence-based review of e-HRM and strategic human resource management. *Human Resource Management Review, 23*(1), 18–36.

Marler, J. H., & Parry, E. (2016). Human resource management, strategic involvement and e-HRM technology. *The International Journal of Human Resource Management, 27*(19), 2233–2253.

Meijerink J, Boons M, Keegan A & Marler J (2021) Algorithmic human resource management: Synthesizing developments and cross-disciplinary insights on digital HRM. The International Journal of Human Resource Management, 0(0):1–18. doi:https://doi.org/10.1080/09585192.2021.1925326

Moher, D., Liberati, A., Tetzlaff, J., Altman, D. G., & The PRISMA Group. (2009). Preferred reporting items for systematic reviews and meta-analyses: The PRISMA statement. *PLoS Medicine, 6*(7), e1000097.

Mozgovoy, V., & Mettler, T. (2019). Internal development as access strategy to information and communication technology in electronic human resource management for sustaining employee well-being. *it - Information Technology, 61*(5–6), 265–272.

Munn, Z., Peters, M. D. J., Stern, C., Tufanaru, C., McArthur, A., & Aromataris, E. (2018). Systematic review or scoping review? Guidance for authors when choosing between a systematic or scoping review approach. *BMC Medical Research Methodology, 18*, 143.

Nura, A. A., & Osman, N. H. (2013). Gauging the effect of performance management and technology based human resource management on employee retention: The perspective of academics in higher educational institutions in Sokoto State Nigeria. *Asian Social Science, 9* (15), 295.

Nurlina, N., Situmorang, J., Akob, M., Quilim, C. A., & Arfah, A. (2020). Influence of e-HRM and human resources service quality on employee performance. *The Journal of Asian Finance, Economics, and Business, 7*(10), 391–399.

Obeidat, S. M. (2016). The link between e-HRM use and HRM effectiveness: An empirical study. *Personnel Review, 45*(6), 1281–1301. https://doi.org/10.1108/PR-04-2015-0111

Oberst, U., De Quintana, M., Del Cerro, S., & Chamarro, A. (2020). Recruiters prefer expert recommendations over digital hiring algorithm: a choice-based conjoint study in a pre-employment screening scenario. *Management Research Review, 44*(4), 2040–8269.

Parker, R., & Bradley, L. (2000). Organisational culture in the public sector: Evidence from six organisations. *International Journal of Public Sector Management, 13*(2), 125–141. https://doi.org/10.1108/09513550010338773

Persson, M., Redmalm, D., & Iversen, C. (2021, Forthcoming). Caregivers' use of robots and effect on work environment – A scoping review. Journal of Technology in Human Services.

Rahman, M., Mordi, C., & Nwagbara, U. (2018). Factors influencing E-HRM implementation in government organisations: Case studies from Bangladesh. *Journal of Enterprise Information Management, 31*(2), 247–275.

Rana, D. (2018). The future of HR in the presence of AI: A conceptual study. *SSRN Electronic Journal*. https://doi.org/10.2139/ssrn.3335670

Ranerup, A., & Henriksen, H. Z. (2019). Value positions viewed through the lens of automated decision-making: The case of social services. *Government Information Quarterly, 36*(4), 101377.

Reddick, C. G. (2009). Human resources information systems in Texas city governments: Scope and perception of its effectiveness. *Public Personnel Management, 38*(4), 19–34.

Reddy, A. J. M., Rani, R., & Chaudhary, V. (2019). Technology for sustainable HRM: an empirical research of health care sector. *International Journal of Innovative Technology and Exploring Engineering, 9*(1), 2919–2924.

Rokeach, M. (1973). *The nature of human values*. Free Press.

Roy, R., & G., & Jegan, P. (2019). E-HRM practices in commercial banks: An empirical study in kanniyakumari district. *International Journal on Emerging Technologies, 10*(3), 408–412.

Schein, E. H. (2017). *Organizational culture and leadership* (5th ed.). Wiley.

Strohmeier, S. (2009). Concepts of e-HRM consequences: A categorisation, review and suggestion. *The International Journal of Human Resource Management, 20*(3), 528–543.

Strohmeier, S. (2020). Smart HRM–a delphi study on the application and consequences of the internet of things in human resource management. *The International Journal of Human Resource Management, 31*(18), 2289–2318.

Sucharew, H., & Macaluso, M. (2019). Progress notes: Methods for research evidence synthesis: The scoping review approach. *Journal of Hospital Medicine, 14*(7), 416–418.

Sundberg, L. (2019). Value positions and relationships in the Swedish Digital Government. *Administrative Sciences, 9*(1).

Tambe, P., Cappelli, P., & Yakubovich, V. (2019). Artificial intelligence in human resources management: Challenges and a path forward. *California Management Review, 61*(4), 15–42.

Tansley, C., Kirk, S., Williams, H., & Barton, H. (2014). Tipping the scales: Ambidexterity practices on e-HRM projects. *Employee Relations, 36*(4), 398–414.

Troshani, I., Jerram, C., & Hill, S. R. (2011). Exploring the public sector adoption of HRIS. *Industrial Management & Data Systems, 111*(3), 470–488.

Turner, A. (2015). Instilling public service values and professionalism through information literacy. *Journal of Public Affairs Education, 21*(1), 41–54. https://doi.org/10.1080/15236803.2015.12001815

Ulrich, D., Younger, J., Brockbank, W., & Ulrich, M. D. (2013). The state of the HR profession. *Human Resource Management, 52*(3), 457–471.

Upadhyay, A. K., & Khandelwal, K. (2018). Applying artificial intelligence: Implications for recruitment. *Strategic HR Review, 17*(5), 255–258. https://doi.org/10.1108/SHR-07-2018-0051

Vallor, S. (2015). Moral deskilling and upskilling in a new machine age: Reflections on the ambiguous future of character. *Philosophy & Technology, 28*(1), 107–124.

Vrontis D, Christofi M, Pereira V, Tarba S, Makrides A & Trichina E (2021) Artificial intelligence, robotics, advanced technologies and human resource management: A systematic review. The International Journal of Human Resource Management, 0(0):1–30. doi:https://doi.org/10.1080/09585192.2020.1871398

Wahyudi, E., & Park, S. M. (2014). Unveiling the value creation process of electronic human resource management: An Indonesian case. *Public Personnel Management, 43*(1), 83–117.

Wirtky, T., Laumer, S., Eckhardt, A., & Weitzel, T. (2016). On the untapped value of e-HRM: A literature review. *Communications of the Association for Information Systems, 38*(1), 2.

Wirtz, B. W., Weyerer, J. C., & Geyer, C. (2019). Artificial intelligence and the public sector—Applications and challenges. *International Journal of Public Administration, 42*(7), 596–615.

Witesman, E., & Walters, L. (2013). Public service values: A new approach to the study of motivation in the public sphere. *Public Administration, 92*(2), 375–405. https://doi.org/10.1111/padm.12000

Zeebaree, S. R., Shukur, H. M., & Hussan, B. K. (2019). Human resource management systems for enterprise organizations: A review. *Periodicals of Engineering and Natural Sciences (PEN), 7*(2), 660–669.

Integration of RPA in Public Services: A Tension Approach to the Case of Income Support in Sweden

Mariana S. Gustafsson

1 Introduction

Scandinavian welfare services and public administration are currently undergoing advanced digitalization, which makes possible introduction of robotic process automation (RPA) and paving the way for different artificial intelligence (AI). The welfare services are steered by political goals for equal access to high-quality services for all members of society and the political ambition is to draw the benefits from technological developments to address such goals and cope with societal challenges in health, social care, and education (Regeringskansliet, 2017; Regeringskansliet & SALAR, 2016).

As its name suggests, RPA relates to involvement of robotic tools in data processing for different purposes, such as optimization of data management for analysis, prediction, and decision-making. Or as Lindgren (2020) describes it, RPA is a software program designed to perform a pre-set number of tasks based on data, features, and rules available in different IT systems. In terms of delimitation of scope and predefinition of tasks, RPA qualifies as "algorithms" (Bannister & Connolly, 2020), a term that will also be used in this chapter. Even though RPA and AI share the same key concepts of nonhuman agency and performativity of human-like intelligence, they are not the same thing, the latter being a much more complex and potent system with more advanced learning-, autonomy-, and decision capabilities (Misuraca & van Noordt, 2020).

Similar to the common promises for technological innovation, the expectations and the policy goals in case of RPA too are associated with greater data efficiency and effective decision processes that would bring more value for the user (Nielsen et al., 2009). Importantly, use of RPA in public services will need to comply with

M. S. Gustafsson (✉)
Department of Management and Engineering, Linköping University, Linköping, Sweden
e-mail: mariana.s.gustafsson@liu.se

G. Juell-Skielse et al. (eds.), *Service Automation in the Public Sector*, Progress in IS,
https://doi.org/10.1007/978-3-030-92644-1_6

democratic demands on enhanced institutional transparency and accountability (Bannister & Connolly, 2020; Misuraca & Viscusi, 2020), but also address risks of data security, individual integrity, algorithm bias, and accountability in case of breach (Bertot et al., 2010; Bonsón et al., 2012; Grimmelikhuijsen & Meijer, 2015).

Introduction of new and advanced technologies in organizations is a challenging process as such (Luna-Reyes, 2017; Meijer & Thaens, 2010; Puron-Cid et al., 2020; van Noordt et al., 2020). When these are introduced in public authorities that are politically steered and governed through a mix of policy making and management structures, such changes become rife with tensions, dilemmas, and failed attempts—that can indicate important moments of learning, innovation, adaptation, or transformation (Bannister & Connolly, 2015, 2020; Gustafsson, 2017; Van Cauter et al., 2015).

States of unrest, pressure, resistance, as well as contestation of meaning guided by different logics—characterize dynamic tension fields that are valuable for deepening the understanding of the process of integration of advanced technologies in human decision-making and services. To understand RPA practices, their common or unique municipal contexts, and the transformations of services that they set out, it is critical to identify and understand the assumptions that underly such arrangements (Whittlestone et al., 2019), the different logics that are at play, as well as the tensions and value conflicts that arise.

In this chapter, I will show that in the context of advancing digitalization of public services, the introduction of RPA-tools in decision processes employed by public servants is a realm rife with tensions, ambiguities, and conflicts that need a careful attention and a thorough understanding. Such an understanding, I argue is critical, as it preconditions and affects how RPA-integration in public service is to be managed and organized. Furthermore, I will argue that the existence of tensions in connection to RPA introduction presents important indicators of the public organizations' capacity and readiness to re-examine its current practices and core services in relation to democratic goals. Beyond the municipal organization, algorithms supported decision-making, require new, cross-jurisdictional governance and coordination mechanisms, as well as new institutional arrangements (Gustafsson, 2017). Such transformations entail, at least a thorough revision, if not totally new logics on service administration, organization management, professional roles, and relation to the clients.

1.1 Purpose and Research Questions

The purpose of this chapter is to uncover and analyze key tensions that relate to governance, management, and organization that accompany the integration of RPA in public services. The tensions will be analyzed in an agency–structure framework (Giddens, 1984; Wendt, 1987) and using empirical data on introduction of RPA-tools in income support services in three Swedish municipalities. The analysis will be guided by two research questions: what tensions underly RPA-integration by actors in their practices; and how these tensions can be understood based on an agency–structure framework.

This chapter proceeds with a brief overview of research on RPA in public organizations and services. Thereafter, a tension approach to RPA-integration, based on agency–structure perspective is presented and operationalized into an analytical model (summarized in Table 1). Then we proceed with specifying the methodological choices as well as the empirical material and the sources used in the analysis. Then follows an analysis of the tensions along the structures identified by the analytical model: governance, public administration, and implementation management. This is followed by an analysis that explains the different tensions based on the agency–structure framework. Finally, we summarize the findings and conclude on critical issues for further research and implications upon theory and practice.

2 Previous Studies

Importantly, to identify and understand critical tensions that underlie automation of decision-making in public services, we need to consider the problems of transparency and accountability that are critical for the governing legitimacy mechanisms. Based on current, still limited research, algorithms are not and cannot be objective, as they are designed by humans who think, act upon, and design in accordance with personal, organizational, professional, and cultural values (Bannister & Connolly, 2020). Opacity of sophisticated RPA-tools and processes goes beyond technical savviness of the users, especially when accessibility to the code is restricted by private or public ownership regulation, the limited access to underlying data, the accuracy of machine learning models, and specific practices generated in the different service and organizational contexts (Engstrom & Ho, 2020). Studies that focus on the issue of transparency and democratic as well as legal oversight have highlighted challenges for the users (including the engineers themselves) in understanding the processes, principles, and values encoded in the algorithms, as well as their alignment with the human rights, democratic, and fairness principles (McGregor et al., 2019). Furthermore, as the algorithmic processes are accepted and used more widely, there arises the risk of trusting the algorithm-generated information as truths, or utterly neutral and correct decisions, may enhance current or generate new structural inequalities and social divides (Favaretto et al., 2019; Ingemarsdotter et al., 2020).

Recent studies that have looked at RPA in specific public services confirmed the problems specified above and found that algorithmic processing of registry data in the case of child benefit services in Norway created a differentiated process for the different citizen groups leading to inequality in provided service quality (Larsson, 2021). A similar study on child welfare data systems in England questions the neutrality of the system as a tool for decision-making and identifies ownership structures, policy agendas, organization practices, and legal frames that influence the design of the data systems (Redden et al., 2020). Citizens' own general negative attitudes to the automation of decision-making by public authorities, based on the Swedish case (Denk et al., 2020), are raising serious concerns regarding legitimacy of RPA-agents and practices in public services.

Another string of literature explaining the challenges of algorithmic or other advanced digital tools found that full automation of decision making, especially those involving discretion based on professional skills and experience, is difficult to design (Busch, 2020), and that it significantly affected bureaucratic decision autonomy (Giritli Nygren et al., 2013). Professional identity and decision complexity were the strongest motives in bureaucrats' reticence toward full automation (Busch, 2019). Such findings are not surprising considering earlier results by Sørensen and Torfing (2011) who emphasized the "mental and organizational silos," as outcomes of institutional and hierarchical steering during decades, while others were concerned about institutional rigidity, formalism, and risk aversion (Cordella & Tempini, 2015; Duivenboden & Thaens, 2008), or dominating service provision logics (Vassilakopoulou et al., 2017) that constrained use of advanced technologies in the public sector.

Gustafsson (2017) has previously problematized the dominance of established managerial and evolutionary approaches and argued that these leave important process-related, subjective individual, and professional aspects out of the analysis of change involved in advanced digitalization of public services. Based on the case of RPA in income support in a Swedish municipality, Gustafsson and Wihlborg (2019) pointed to tensions that arose from different logics that drove the different professions in the practice of case management and service provision: unit management, caseworkers, IT strategists, and higher-level functionaries. In earlier analyses, Gustafsson (2017) identified and explained four tension fields: governance, organizational, professional, and individual levels in the context of advancing digitalization in public administration. Joining the call for further research, in terms of more empirical data and relevant theoretical models, the author—by taking a tension approach and based on three case studies, hopes to bring further insight on the change dynamics of integration of advanced technologies in decision-making and quality of public services.

3 A Tension Approach in an Agency–Structure Framework to RPA

A central theme preoccupying academia, politicians, and public servants alike is whether local governments are equipped to use advanced digital technologies to provide high-quality public services in line with modernization policies, democratic values, and citizen expectations. One way to approach this theme is to pay attention to the tensions that arise when existing practices, information infrastructures, and norms are challenged by new technologies, thus creating a propensity for change, adaptation, and innovation (Bertot et al., 2008; Ritala et al., 2017; Savoldelli et al., 2014; Wagenaar, 2006). Tensions in the social context are characterized in terms of states of unrest pressure, resistance, contestation of meaning, conflicts or dilemmas, but also the need for learning, action, creativity, leadership, decisions, and change in the organization. Thus, such tensions are states of latent striving, unrest, or pressure in agency–structure dialectics that develop toward stability or change (Gustafsson, 2017).

The assumption is that introduction of new technologies in practice is a process that involves a multitude of actors with different, limited, or even no understanding of the specific problems that arise in practice in the interface between the public service providers and the citizen, the norms and regulations that are translated by different professions when applied in practice. Such a multitude of understandings and logics involves a realm of tension when they need to collaborate to learn, understand, and use new technologies in their practices. Thus, tensions emerge when new practices, in our case through the use of advanced technologies, become part of an individual's sense-making of social, professional, and organizational reality (Berger & Luckmann, 1991; Searle, 1996). In the context of organizing activities, tensions have been earlier studied, especially in terms of strategic dualities and contradictions (Karlsson & Montin, 2012; Ritala et al., 2017; Smith & Lewis, 2011).

Another assumption is that technologies are not neutral tools (Winner, 1977, 1980). The more advanced the technologies are, especially when it comes to algorithms, different RPAs and AI, at least at the moment of writing, the more black boxes they present for most users in public administrations, politicians, and citizens. Algorithm bias and accountability issues connected to them show that their adoption and use in public services raises numerous questions that generate tensions, resistance, and hesitance in practice.

Elsewhere, I showed that different logics of implementation of e-government in governance cooperation structures made up a tension-laden context in local public administration (Gustafsson, 2017). For example, uncertainty about rising volumes of information and its heterogenous character generated by new technologies emerged in a context of unclear organizational arrangements and institutional ambiguities (Gustafsson, 2017). Professionals' use of the advanced digital platforms was also rife with hesitance, resistance, but also curiosity, where the perceptions and the subjective meanings of security were undergoing change (Gustafsson, 2017). However, users' trust and accountability norms seemed to alleviate some of these tensions. In turn, such tensions involved pressures on trust mechanisms, both internally in public administration and in relation with the citizens. Therefore, it was concluded that depending on how actors in public administration coped with such tensions in practice, citizens' trust for them would be affected (Gustafsson, 2017).

3.1 The Agency–Structure Problem

An agency–structure frame of analysis can provide useful analytical tools for studying the RPA introduction in public administration of welfare services. In its turn, the case of RPA (and subsequently AI) can inform the theory by involving new opportunities or risks for the agency, or even introducing hybrid or nonhuman agency into the focus. The agency–structure framework builds on two assumptions of social organization: Humans and their organizations are intentional actors that form society through their actions; These actors establish relationships that make up the structure of society (Dawe, 1978; Wendt, 1987). Based on these two assumptions, different theories have strived to explain the relationship between these agents and their role in shaping social organization (Giddens, 1984; Wendt, 1987).

Different approaches to the problem have generated conceptualizations based on individualism, structuralism, and structurationism (Cetina & Cicourel, 2014; Wendt, 1987). While the first two attribute critical importance to either agents or structures, the structuration theory considers both agents and structures as equally critical for the organization of society, or as Wendt put it, having "equal ontological status" (Wendt, 1987). While both agents and structures are made part of the same ontological entities, their properties, behaviors, and effects shall be understood in relation to each other, as they are co-dependent and co-consistent in the social order.

But, as complex technologic infrastructures, as well as algorithmic agency, are increasingly made part of the social structures in society, the duality of the "agency–structure" relationship is challenged, which requires a re-examination of the agency–structure problem in social sciences. Furthermore, in line with the purpose of the paper, I argue that if advanced technologies are gaining agency, as is possibly the case of RPA and become autonomous (as is the case of AI), and not only integrate or mediate in the social structures, but will also shape them—this will generate fields of tension, conflicts, and paradoxes that will also affect the social agency, structure, and relationships of social organization, possibly fundamentally transform them. Therefore, a tension approach to the study of agency and structure is a useful way to identify processes of integration, adaptation, or transformation of social organization.

3.2 Structuration Theory, Technology Agency, and Its Critique

Structuration theory's core concepts are: agency, social structure, social system, the duality of structure, time-space extension, and institutional order, some of which will be used in this chapter analysis (Giddens, 1984; Thrift, 1985). According to Giddens, agency is a series of repetitive practices of reflexive actors, drawing upon knowledge from conscious and unconscious mind, as well as constant reflexivity upon their actions. One can reason in the light of the central theme of this volume, that if advanced RPA, and eventually AI, shall account for a knowledge and reflexivity base stretching beyond an individual human actor's analytical capability, then it most likely can enact some sort of agency and probably a powerful such. The conceptualization of agency in structuration theory (Giddens, 1984), as presenting repetitive and routine actions of social integration, fits therefore the current capabilities of RPA (at least technically and operationally, as I will show in this chapter). However, this conceptualization has been criticized for not focusing enough on the reflexivity and the creativity of human agency (Thrift, 1985).

Social structure consists of rules and resources that are stocked in knowledge upon which agents act in day-to-day practices. The pattern of these agent-driven practices is generally reproductive, is situated in specific contexts, happens at different scales and with different degrees of order, thus making up for the social system (Thrift, 1985). The duality of structure means that agents and structure are mutually producing, changing, and shaping each other and that the distinction between them is analytical, and not ontological (Hay, 2002; Marsh, 2018).

New technologies, in this theory, have the role of intermediating system integration, thus replacing face-to-face interaction among agents, and supporting the system's viability and extension over time. In theory, the issue of human and nonhuman agency and their role in sustaining stability or driving change of the social system—presents a fundamental tension that calls for reconsideration of the theory itself in the light of current technological development that both enables and limits human agency on the one side, and gains agency of its own on the other.

Even though routine and repetitive patterns are necessary for system integration, the theory was criticized for perceiving processes of social organization as too orderly, friction- or conflict free (Thrift, 1985). Although the theory focuses on agency, another important critique is the lack of attention to the institutions as critical agents in integration of social systems (Thrift, 1985). Partly, in line with this critique, but recognizing the duality of agency and structure, I suggest reconsidering it in the light of technologies increasing agency, the creative reflexivity of human agency, and the messy processes of social change.

3.3 Tension Analysis Model

Building on structuration theory, I propose a framework for the identification and analysis of tensions that may underly the introduction of RPA in public services in practice. The analytical model and its operationalization targeting the potential tensions are summarized, with some examples in Table 1 and shortly described below.

Table 1 Agency–structure for RPA introduction and integration

Structure/Agency	Local government & municipal administrations	Professions and street—level public servants	Citizens—final beneficiaries of welfare services
Governance: politics, policy, legislation, norms and principles	Lack of support in legislation Gray areas in legislation Lack of policies, unclear political priorities and support, central vs local, large vs small actors	Lack of support in legislation Gray areas Lack of operationalization of laws into praxis	Accountability, Transparency, Trust—difficulties Equal access and Inclusion—digital divides/ diversity
Public Administration: legislation, norms and principles	Lack of support in legislation Gray areas New Public Management	Single profession units vs mixed professions units: different logics Ethics of care vs ethics of justice	Accessibility, Competence Fairness vs digital divides/ diversity
Public organization/ Management	Political committees vs administrations Unclear leadership Efficiency vs quality of service	Subordination to different committees: different priorities and logics Leadership vs case managers Competences	Trust vs control
Technical infrastructure	Registries and databases: capacity versus need Data Operative systems Service hubs Robot factories	Individual case handling vs standardization Efficiency and quality Algorithm bias	Secure and simple to use services Accessibility, Competence Fairness vs digital divides/ diversity
Data	Data security Data use	Client data security Data quality Sensitive data	Personal data security

To briefly clarify the model and the empirically based examples (based on the Swedish case) in Table 1, RPA introduction in public services will include actors such as local governments with their municipal administrations, street-level public servants, and their professional backgrounds, as well as citizens. However, the extended agency includes central government with its agencies, private companies, and public–private partnerships involved in production or support of data registry, RPA- and other technologies, and society at large. These reflexive actors act upon individual and collective knowledge stocked in governing norms, rules, and principles such as institutions, legislation, policies through iterative practices. In this model, Technology and Data are the critical elements of both agency and structure—as they present the medium through which reflexive agency is engaging in practices, which also become part of the structure. We can thus hypothesize that the more data, technology, and knowledge is advancing—the more advanced agency and structures will be developed, which implies more intricate and rifer with tensions the agency—practice—structure relation will become.

Applied to the case of RPA-integration in public services, frictions, contradictions, or ambiguities may arise in governance structures such as gray areas in legislation that concern handling of complex or sensitive data and technologies in public administration in relation to the principles of "public record"; local authorities' autonomy to organize and deliver services versus central authorities control mechanisms; lack of- or inconsistent political support, ambiguous policy priorities concerning automation or difficulties of applying policy into practice by the service units. For the citizen agency, values such as transparency and accountability of algorithmic decisions, handling of personal data, accessibility of services can create difficulties, misunderstandings, or knowledge that empower or undermine—trust and legitimacy in public services. These are only a few examples that are ordered in Table 1 according to the proposed framework.

In the public administration structures, the emergence of new functions, new knowledge and competence needs, and reallocation of competences creates—through different forms of protest, questioning, re-skilling—a new composition of service units, from one dominant profession and ethics, toward integration of different professions and different ethics, from downpipe organization and collaboration on services toward a cross-organization and cross-jurisdictional one (Gustafsson, 2017). For the relation citizen—administration, a change in administrative practices can also involve releasing control and relying on trust, or a re-allocation of control mechanisms now mediated by more technology, data, and knowledge. Thus the subordination of the public service units to the different political committees that have different political agendas, engagement, and priorities would present important conditions for RPA integration and its effects upon both agents and structures.

4 Methods and Material

Management and evolutionary perspectives dominate the research on advanced e-Government (Gustafsson, 2017), which raises the need for interpretive perspectives and thorough empirical cases on automatization in local public administration (Bannister, 2010; Meijer & Bekkers, 2015). This analysis takes an interpretive approach, focusing on practitioners and professional sense making, anchored in their experience and value-based subjectivity and inter-subjectivity (Alvesson & Sköldberg, 2009).

The empirical material is based on field notes from 7 observations in situ in two of the municipalities, notes from meetings and interviews (25 posts) with unit managers, middle managers, as well as group leaders and social workers. In addition, internal documentation (12 posts) on planning, process assessments, strategies, organization charts, and action plans, have complemented the primary sources.

Initially, the material was collected inductively, being guided by the intention to grasp the practices of implementation of RPA, which subsequently switched to an abductive process (Alvesson & Kärreman, 2011; Alvesson & Sköldberg, 2009) by using a theoretical lens focusing on tensions in actor–structure relationships.

The qualitative analysis followed a pattern-seeking and coding logic, employing theoretically informed categories such as "actors," "governance," "administrative organization," "management" (Table 1). The analysis followed a dialectics between patterns and fragmentation, where "the non-obviousness of meaning as well as the potential of multiple meanings" that set up for ambiguities was focused and examined (Alvesson & Kärreman, 2011).

Finally, it is important to note the limited number of actors and municipalities examined in this study poses certain limitations upon generalizability of the findings to a larger group of cases. The findings and conclusions should therefore be considered carefully and complemented with further or similar studies on processes of RPA-integration. However, considering the fact that two pioneering municipalities and one municipality are focused here, their experience, methods, and contexts are nevertheless valuable for understanding processes of stability and change in agency–structure relationship, when advanced technologies are integrated into public administration and services.

4.1 RPA in Three Swedish Municipalities

Income Support units, in three different municipalities currently implementing or preparing for RPA were the focus of the empirical research. The municipalities—anonymized here as "M," "N1," and "N2," lie in the southeastern part of Sweden, differ in size and socio-economic conditions. According to SALAR[1] classification (2017), our sample includes one medium-size municipality, one rural municipality, and one large municipality. One municipality is currently advancing their digital services and prepare for RPA-introduction (M), while the two others (N1 and N2) have introduced RPA since 2017, being among the five front-runners, and are both working on their respective second versions of RPA-tools. RPAs in the case of N1 and N2 involved a software that was calculating the amount of income support according to pre-defined national norms and issued a decision to the client[2] when the case was straightforward. When the case deviated from the rules, RPA was replaced by a human caseworker. In the case of N1—a registry of deviating cases was carefully stored and analyzed, with a purpose of re-designing the RPA. Importantly, the RPAs in N1 and N2, was part of a larger process of the income support service, with human handling of registration of new cases, human supervision and control of deviations and routine checks of RPA generated decisions, human handling of

[1] SALAR: Swedish Association of Local Authorities and Regions.

[2] Applicants to income support benefits.

personal contact with the client that provided coaching and support. RPA in M was rudimentary in the form of an operative system that used automated registry control, and human handling of cases was to a large extent manual.

5 Tensions in Governance

Decision automation in public services is being implemented in Sweden since the beginning of 2000s, with subsequent modification to the laws (Prop 2003/04:145). Its explicit introduction in the Administrative Procedure Act (SFS 2017:900) allowed for general use of automated decisions in the administrative practices, inscribing the same responsibilities, accountability, and documentation of the decision maker, be it human or automatic agent (Prop. 2016/17:180).

While these modifications set out automated decision-making in state public administration, such as The Swedish Tax Agency, Försäkringskassan (Social insurance agency) and the Swedish Transport Agency, for the local authorities, and especially for the social services, the legal support for automation of decision functions was lacking from the Local Government Act (SFS 2017:725) and from the Social Services Act (SFS 2001:453). Both regulated only human decision-making. Confusion on the legal grounds of excluding the human from the exercise of authority, also emerged in relation to the nature of the decisions of the local authorities. In contrast to the decision authority in state administration, the decision-making in Swedish local authorities is a collective undertaking by the elected committees, which is in practice delegated to the professional public servants (SFS 1974:152).

But the legislative support for RPA was interpreted differently in practice and has led a few front-runner municipalities to launch whole or partial automation of case handling in Income support services. Others have questioned the legality of RPA and SALAR has called for more clarity or change in the respective legislative acts. In response, a governmental public investigation has recently proposed to modify the respective acts in order to enable delegation of decisions on certain cases from the respective service committee to an automated decision function (SOU 2021:16). This different interpretation of the legal support for RPA clearly shown in the three studied municipalities. The front-running ones, N1 and N2 set out early in 2017, with either strong support from the administrative and political management or due to a large scope of action delegated to the service unit management. M did not have either of these, but instead a strong reluctance came from senior social worker professionals in the unit.

Fragmented steering and coordination of digitalization, especially in relation to vital digital infrastructure and information security, among local authorities was another important ground for tensions (2016/17:54; Ingemarsdotter et al., 2020). This happens in the context of a decentralized public administration with long stretching autonomous powers of local governments, thus making national coordination and steering of digital infrastructures that pre-condition RPA-use in municipal services in a governance dilemma. Especially small, or municipalities with sparsely or decreasing population were found vulnerable in terms of infrastructure and competence resources, risk-taking, and direction for total digitalization of services, at the same time as they needed to sustain access to high quality of services for the inhabitants and attract companies (Gustafsson, 2017; Syssner, 2014). Municipality M seemed to face such a dilemma in this case.

6 Tensions in Public Administration

The standardization logic versus individualization logic of decisions in social services present another challenging aspect affecting RPA-integration. Values such as efficiency, productivity, and measurability of quality goals in public services that underly new public management structures in Swedish public administration build on a logic of standardization of case management and decisions (Hasselblad & Sundberg, 2020; Pierre, 2016; Ranerup et al., 2016). Advancing digitalization, involving automation, furthermore embeds such a logic in the administrative structures of public service provision (Cordella & Paletti, 2018).

The individualization principle of decisions in case management in income support according to the Social Services Act, meaning that each decision on the case is taken based on careful assessment of the entire household's situation and the individual's ability to work, seemed to clash with the logic of standardized management and decision processes that are required in an eventual automation (Gustafsson & Wihlborg, 2019). However, as our empirical material showed, this clash of logics involved different actions for the different municipalities—hindering some to launch in automation initiatives (Municipality M), while prompting others (Municipalities N1 and N2) to look over, rethink, rebuild, and adapt their case processing, decision-making to automate them wholly or partly.

Our material also shows that RPA-integration has actualized the question of control versus trust in the relation between the street-level public servant and the citizen. Following the individualization logic by which the decision on cases was based on the rigorous 'manual' control of the records and activities of the clients (adminstrative term for citizens using the servicies), the three municipalities in our case have reconsidered releasing control mechanisms or rely on trust in their clients. For Municipality M, automation brought the issue of exercise of control to the fore, and raised concerns on wrong decisions and incorrect benefit payments, making it a crucial issue of tension and resistance initially. For the Municipality N2, the issue of control was addressed prior to automation, in the context of migration to another committee and reorganization of the work routines and priorities. While the Municipality N1 has reorganized their control and trust mechanisms by reinventing the routines and reallocating of competences through integration of the RPA in the case processing. Their reorganization and the new work structure have integrated both control and trust for the benefit of the legality of their decision and the increased trust of the clients.

Yet another ground for tensions seemed to be lying in the organization of the income support under specific committees guided by different political priorities. Those service units that obeyed under the Social Services Committee were clearly guided by an ethics of care logics and employed mainly social workers as street-level public servants (Municipalities M and N1). The unit in the Municipality N2 was moved under the Work and Entrepreneurship Committee, where the "primacy of work" (*Arbetslinjen*) was the guiding principle, and which has employed a diversity of professionals. Clashes and resistance arose in the latter case when the different logics replaced each other and most of the social worker professionals quit their jobs in protest as in Municipality N2, but also in other municipalities not part of our case sample (Bolin & Loth, 2018).

7 Tensions in Change Management and Implementation

RPA-integration in practice has been understood as a complex organizational change happening in the context of modernization of service delivery, and availability of new technological capabilities and growing volumes of data in public administration (Ingemarsdotter et al., 2020). Managing RPA introduction and its integration in the decision process has involved for the service units in our sample to introduce functional changes to the operative systems on case management or changing them totally, create new work routines, and reallocate tasks in the teams—mean while they needed to keep up with the ongoing casework.

A gap was observed between expectations of political representatives in the political committees, expressed in single phrases such as "*we shall automatize our processes,*" and the reality of risk taking, managing, and implementing the changes in practice by the service units. In Municipality M such a gap has complicated and delayed the planning for the unit management. For the other two units (Municipalities N1 and N2), RPA integration involved understanding of the mechanisms and rebuilding of RPA-tools to fit the units' needs, and its functionality in the decision chain on the cases, instating a new function that integrates the RPA, mapping of internal methods, involving and investing in the right competences, time management, and risk taking.

Related to the gap between expectations and actual capabilities was the committees' high expectations of quality and efficiency of RPA versus the digital infrastructure readiness and the dialogue between the different support units inside the municipal organizational and the externally contracted RPA-developers. A productive dialogue between the implementing units and the municipal IT support units was crucial in the RPA-integration process for all three units in our focus. In the municipalities M and N2, the technical support function was only partially involved in the dialogue with the development, and was not always the kind of technical, explanatory, or problem-solving support that the unit managers have hoped for and needed.

8 Governance, Organizational, and Management of RPA in Public Services

Rule of law provides the fundamental structural base upon which the municipalities exercise their authority and administer the welfare services. As our examples show, the municipalities and their administrations act in the same governance structure but choose to enact their agency differently. When the respective legal base is not clear enough, there is room for interpretation that leads to different action paths. While some municipalities considered the Administrative Procedure Act (SFS 2017:900) as sufficient to introduce automated decision-making, others have questioned the legality by referring to the complementary laws (The Local Government Act and the Social Services Act) that were central too and that did not provide the ground for delegation of decision agency to automated functions.

It is important to understand how such a different outcome of the law interpretation was possible and how differently the respective municipalities exercised their agency. Provided that at least one of the legal bodies allowed for the general use of automation and that there existed other recurring practices of the pioneering state agencies that successfully introduced automation, in addition to the political will, it was a solid enough ground for the two front-running to launch the automation changes. The same ground, however, was not sufficient for the third municipality in our case, which gave heavier importance to the other two laws and lacked political and technical support.

Involving a process of deep digitalization, automation raises tensions connected to fragmented coordination and unclear steering in practice of vital digital infrastructure and information security. The three focused municipalities had different resource bases to integrate automation—technical, infrastructure, and special competence—to result in either copying a ready-made RPA-solution, design an own RPA-tool or just to add some additional functions to an existing operative system. These three different bases show evidence already now that they will eventually lead to three different paths and automation practices. However, they acknowledge the need for a common coordination mechanism to ensure that these paths are somewhat similar, be it partly or entirely automatized decision processes, but with a comparable output for the client in terms of quality, security, and the rule of law.

The decentralized structure of the welfare system, underpinned by strong self-determination principles creates powerful agency at the local level of government. But in practice, and over time, the 290 Swedish municipalities may develop differing agencies and affordances for advanced technologies in public services. Common coordination mechanisms, the role of the central government and the common secure infrastructure based on advanced technologies—present currently an intrinsic tensed field of differently potent agencies where capacity, knowledge of- and willingness to integrate technologies seems to play a crucial role.

As our examples show, the municipal units delivering income support services need to grapple with two different logics that steer service provision: the standardization of processes versus the individual assessments in case management.

Importantly, individualization logic is not critical for all welfare services. For example, in building and housing services, in urban infrastructure, in transportation, in tax administration—services where decisions are taken based on general principles, quantifiable parameters, and building on technical, registry, or other numerical data—standardizing methods and tools seem to be more applicable and effective.

While in health care and social care, individualization of decisions on cases is critical as these services are attending to clients in most vulnerable, unique, and complicated health or socio-economic conditions. Professions like social work or psychologist, put the individual and their needs at the center of attention, build a trustful relation that they rely on to help rehabilitation and inclusion in society. Some of these professionals in the public service, in our sample too, have deep concerns facing standardization pressures (along with rationalization and efficiency goals) in relation to the mission and the promise of the welfare state to its citizens. The pressure of standardization and the concerns were experienced most stringently in the context of economic saving demands.

However, it is possible to overcome this tension and accommodate the two logics, as was proved by one of the units (Municipality N1) in our sample. When designing their RPA, they could map, assess, and streamline their work process, in such a way as to find appropriate functionality and utility for the RPA-tool in the decision chain in case management, to delimit the its agency and to introduce thresholds that were crucial for human decision. The work teams were re-organized to deal on the one side with the administrative part involving RPA and on the other side to focus on the coaching part involving intensive and recurrent human interaction with the clients. They have thus managed to both standardize and automatize a part of the administrative process and reallocate competence and more time for human interaction. Based on the evidence from our empirical data, we can see how the service units exercise different agency when addressing the tension between standardization and individualization of decision making on cases.

9 Concluding Remarks and Further Research

The municipalities, with the Income support units in our focus, faced the issue of introducing RPA to streamline their decision processes on cases. Based on our findings, it is no doubt that introduction of RPA in the public service raised fundamental questions concerning governance principles, the mechanisms of public administration, as well as the management and the organization of work. The presence of multiple tensions shows a reality of reflexive actors who are both acting within a robust structure of public administration and at the same time are changing it through their choices and practices that now include RPA.

In terms of the agent–structure problem, the case of automation in Income support shows that transformation actually was driven both by actors, who changed their practices by integrating RPA and that in turn, showing positive effects, are changing the structure both at fundamental governance level and in organization and management. As we have seen, in a decentralized governance system, actors differ at the moment in their choices and outcomes of their organizational change involving RPA, which also shows how differently they enact their current agency and how ready they are to augment it by using advanced technologies.

Base on this analysis, four important questions that need further attention are: whether, or actually when will the municipalities develop different agencies that will affect the quality and the outcomes of public services; how will governance structure change in order to balance the different municipal agencies to fulfill their democratic mission; how to sustain democratic values such as trust versus control, inclusion versus diversity of needs; and finally how do we understand and explain both in research and in practice the increasing complexity of agency and structure when advanced technology is deeply integrated.

References

2016/17:54, S. *Riksrevisionens rapport om den offentliga förvaltningens digitalisering*.

Alvesson, M., & Kärreman, D. (2011). *Qualitative research and theory development: Mystery as method*. Sage Publications.

Alvesson, M., & Sköldberg, K. (2009). *Reflexive methodology: New vistas for qualitative research*. SAGE.

Bannister, F. (2010). Deep e-government: Beneath the carapace. In H. J. Scholl (Ed.), *E-government: Information, technology, and transformation* (Vol. 17, pp. 33–51). M.E. Sharpe.

Bannister, F., & Connolly, R. (2015). The great theory hunt: Does e-government really have a problem? *Government Information Quarterly, 32*(1), 1–11. https://doi.org/10.1016/j.giq.2014.10.003

Bannister, F., & Connolly, R. (2020). Administration by algorithm: A risk management framework. *Information Polity, 25*(4), 471–490. https://doi.org/10.3233/ip-200249

Berger, P. L., & Luckmann, T. (1991). *The social construction of reality: A treatise in the sociology of knowledge*. Penguin.

Bertot, J. C., Jaeger, P. T., & Grimes, J. M. (2010). Using ICTs to create a culture of transparency: E-government and social media as openness and anti-corruption tools for societies. *Government Information Quarterly, 27*(3), 264–271.

Bertot, J. C., Jaeger, P. T., & McClure, C. R. (2008). *Citizen-centered e-government services: Benefits, costs, and research needs*. Paper presented at the Proceedings of the 2008 international conference on Digital government research, Montreal, Canada.

Bolin, M., & Loth, A. (2018). 12 av 16 socialsekreterare säger upp sig i protest. *SVT Nyheter*. Retrieved from https://www.svt.se/nyheter/lokalt/halland/robot-tar-over-i-kungsbacka

Bonsón, E., Torres, L., Royo, S., & Flores, F. (2012). Local e-government 2.0: Social media and corporate transparency in municipalities. *Government Information Quarterly, 29*(2), 123–132. Retrieved from http://www.sciencedirect.com/science/article/pii/S0740624X1200010X

Busch, P. A. (2019). *Digital discretion acceptance and impact in street-level bureaucracy*. (PhD Doctoral Dissertation). Universitetet i Agder, Agder.

Busch, P. A. (2020). Crafting or mass-producing decisions: Technology as professional or managerial imperative in public policy implementation. *Information Polity, 25*, 111–128. https://doi.org/10.3233/IP-190163

Cetina, K. K., & Cicourel, A. V. (2014). *Advances in social theory and methodology (RLE social theory): Toward an integration of micro-and macro-sociologies*. Routledge.

Cordella, A., & Paletti, A. (2018). ICTs and value creation in public sector: Manufacturing logic vs service logic. *Information Polity, 23*(2), 125–141. https://doi.org/10.3233/IP-170061

Cordella, A., & Tempini, N. (2015). E-government and organizational change: Reappraising the role of ICT and bureaucracy in public service delivery. *Government Information Quarterly, 32*(3), 279–286.

Dawe, A. (1978). Theories of social action. *A history of sociological analysis, 362417.*

Denk, T., Hedström, K., Karlsson, F., & Wihlborg, E. (2020). Medborgarnas inställning till automatiserat beslutsfattning. In *Digitala är vi allihopa?* (pp. 79–89). SOM-institutet, Göteborgs universitet.

Duivenboden, H. V., & Thaens, M. (2008). ICT-driven innovation and the culture of public administration: A contradiction in terms? *Information Polity, 13*(3–4), 213–232.

Engstrom, D. F., & Ho, D. E. (2020). Algorithmic accountability in the administrative state. *Yale Journal on Regulation, 37*(3), 800–854. Retrieved from <Go to ISI>://WOS:000567989300001.

Favaretto, M., De Clercq, E., & Elger, B. S. (2019). Big data and discrimination: Perils, promises and solutions. A systematic review. *Journal of Big Data, 6*(1), 12. https://doi.org/10.1186/s40537-019-0177-4

Giddens, A. (1984). *The constitution of society: Outline of the theory of structuration.* Polity Press.

Giritli Nygren, K., Axelsson, K., & Melin, U. (2013). Public e-services from inside: A case study on technology's influence on work conditions in a government agency. Retrieved from https://lt.ltag.bibl.liu.se/login?url=http://search.ebscohost.com/login.aspx?direct=true&db=edsswe&AN=edsswe.oai.DiVA.org.liu.98041&site=eds-live

Grimmelikhuijsen, S. G., & Meijer, A. J. (2015). Does Twitter increase perceived police legitimacy? *Public Administration Review, 75*(4), 598–607. Retrieved from <Go to ISI>://WOS:000356975600014.

Gustafsson, M. (2017). *Reassembling local e-government: A study of actors' translations of digitalisation in public administration.* (718 Doctoral thesis, comprehensive summary). Linköping University Electronic Press, Linköping.

Gustafsson, M., & Wihlborg, E. (2019). 'It is always an individual assessment': A case study on challenges of automation of income support services. In I. Lindgren, M. Janssen, H. Lee, A. Polini, M. P. Rodríguez Bolívar, H. J. Scholl, & E. Tambouris (Eds.), *Electronic government. EGOV 2019. Lecture Notes in Computer Science* (Vol. 11685, pp. 45–56). Springer International Publishing.

Hasselblad, A., & Sundberg, L. (2020, 31 August–2 September). *When worlds collide: Comparing the logic of the industrial and welfare societies.* Paper presented at the Electronic Government, E-Democracy and Open Government, and Electronic Participation: Joint Proceedings of Ongoing Research and Projects of IFIP WG 8.5 EGOV-CeDEM-ePart 2020, Linköping, Sweden.

Hay, C. (2002). *Political analysis: A critical introduction.* Palgrave.

Ingemarsdotter, J., Eidenskog, D., & Hedtjärn Swaling, V. (2020). *Vilse i lasagnen?– En upptäcktsfärd i den svenska digitaliseringens mångbottnade problemstruktur / Lost in the Lasagna?– A journey into the multi-layered problem structure of Swedish digitalisation.* Retrieved from

Karlsson, D., & Montin, S. (2012). Solving municipal paradoxes: Challenges for Swedish local democracy. *Panorama: Insights into Asian and European Affairs, 2*, 125–136.

Larsson, K. K. (2021). Digitization or equality: When government automation covers some, but not all citizens. *Government Information Quarterly, 38*(1), 101547. https://doi.org/10.1016/j.giq.2020.101547

Lindgren, I. (2020, 31 August–2 September). *Exploring the use of robotic process automation in local government.* Paper presented at the Electronic Government, E-Democracy and Open Government, and Electronic Participation: Joint Proceedings of Ongoing Research and Projects of IFIP WG 8.5 EGOV-CeDEM-ePart 2020, Linköping, Sweden.

Luna-Reyes, L. F. (2017). Opportunities and challenges for digital governance in a world of digital participation. *Information Polity: The International Journal of Government & Democracy in the Information Age, 22*(2/3), 197–205.

Marsh, D. (2018). Meta-theoretical issues. In V. Lowndes, D. Marsh, & G. Stoker (Eds.), *Theory and methods in political science*. Palgrave.

McGregor, L., Murray, D., & Ng, V. (2019). International human rights law as a framework for algorithmic accountability. *International and Comparative Law Quarterly, 68*(2), 309–343. https://doi.org/10.1017/S0020589319000046

Meijer, A., & Bekkers, V. (2015). A metatheory of e-government: Creating some order in a fragmented research field. *Government Information Quarterly, 32*(3), 237–245. Retrieved from http://www.sciencedirect.com/science/article/pii/S0740624X15000568

Meijer, A., & Thaens, M. (2010). Alignment 2.0: Strategic use of new internet technologies in government. *Government Information Quarterly, 27*(2), 113–121.

Misuraca, G., & van Noordt, C. (2020). *Overview of the use and impact of AI in public services in the EU, EUR 30255 EN. Publications Office of the European Union, Luxembourg (2020)* (9276195408). Retrieved from

Misuraca, G., & Viscusi, G. (2020, 31 August–2 September). *AI-enabled innovation in the public sector: A framework for digital governance and resilience.* Paper presented at the Electronic Government: Proceedings of the 19th IFIP WG 8.5 International Conference, EGOV 2020, Linköping, Sweden.

Nielsen, J. A., Medaglia, R., & Andersen, K. N. (2009). *Time to give in: Firm belief in administrative reform as a driver for delivering benefits from IT.* Paper presented at the Electronic Government: Proceedings of the 8th [IFIP WG 8.5] International Conference, EGOV 2009, Linz, Austria.

Pierre, J. (Ed.). (2016). *The Oxford handbook of Swedish politics*. Oxford University Press.

Prop 2003/04:145. *Trängselskatt [Congestion tax].*

Prop. 2016/17:180. *En modern och rättssäker förvaltning – ny förvaltningslag [A modern and legally secure administration - A new administrative procedure act].*

Puron-Cid, G., Gil-Garcia, J. R., & Luna-Reyes, L. F. (2020). Opportunities and challenges of policy informatics: Tackling complex problems through the combination of open data, technology and analytics. In M. Khosrow-Pour, S. Clarke, M. E. Jennex, & A.-V. Anttiroiko (Eds.), *Open government: Concepts, methodologies, tools, and applications* (pp. 236–258). IGI Global.

Ranerup, A., Henriksen, H. Z., & Hedman, J. (2016). An analysis of business models in Public Service Platforms. *Government Information Quarterly, 33*(1), 6–14. https://doi.org/10.1016/j.giq.2016.01.010

Redden, J., Dencik, L., & Warne, H. (2020). Datafied child welfare services: Unpacking politics, economics and power. *Policy Studies, 41*(5), 507–526. https://doi.org/10.1080/01442872.2020.1724928

Regeringskansliet. (2017). För ett hållbart digitaliserat Sverige – en digitaliseringsstrategi.

Regeringskansliet, & SALAR. (2016). *Vision e-hälsa 2025 - gemensamma utgångspunkter för digitalisering i socialtjänst och hälso- och sjukvård.* Retrieved from http://www.regeringen.se/contentassets/79df147f5b194554bf401dd88e89b791/vision-e-halsa-2025.pdf

Ritala, P., Huizingh, E., Almpanopoulou, A., & Wijbenga, P. (2017). Tensions in R&D networks: Implications for knowledge search and integration. *Technological Forecasting and Social Change, In Press, Corrected Proof.* Retrieved from http://www.sciencedirect.com/science/article/pii/S0040162517300094

Savoldelli, A., Codagnone, C., & Misuraca, G. (2014). Understanding the e-government paradox: Learning from literature and practice on barriers to adoption. *Government Information Quarterly, 31*, S63–S71. Retrieved from <Go to ISI>://WOS:000340323800007.

Searle, J. R. (1996). *The construction of social reality*. Penguin.

SFS 1974:152. *Regeringsformen [Instrument of Government].*

SFS 2001:453. *Socialtjänstlag [Social Services Act].*

SFS 2017:725. *Kommunallag [Local Government Act].*

SFS 2017:900. *Förvaltningslag [Administrative Procedure Act].*

Smith, W. K., & Lewis, M. W. (2011). Toward a theory of paradox: A dynamic equilibrium model of organizing. *Academy of Management Review, 36*(2), 381–403. Retrieved from https://login.e.

bibl.liu.se/login?url=https://search.ebscohost.com/login.aspx?direct=true&AuthType=ip, uid&db=buh&AN=59330958&lang=sv&site=eds-live&scope=site.
Sørensen, E., & Torfing, J. (2011). Enhancing collaborative innovation in the public sector. *Administration and Society, 43*(8), 842–868. Retrieved from http://www.scopus.com/inward/record.url?eid=2-s2.0-84855274288&partnerID=40&md5=899edd754434fc0b9fd3caca03a6c1f6
SOU 2021:16. *En väl fungerande ordning för val och beslutsfattande i kommuner och regioner*. Retrieved from Stockholm.
Syssner, J. (2014). *Politik för kommuner som krymper* (Vol. 4). Linköping University Electronic Press.
Thrift, N. (1985). Bear and mouse or bear and tree? Anthony Giddenss reconstitution of social theory. *Sociology, 19*(4), 609–623.
Van Cauter, L., Bannister, F., Crompvoets, J., & Snoeck, M. (2015). When innovation stumbles: Applying Sauer's failure model to the Flemish road signs database project. *International Journal of Public Administration in the Digital Age (IJPADA), pending*.
van Noordt, C., Medaglia, R., & Misuraca, G. (2020, 31 August–2 September). *Stimulating the uptake of AI in public administrations: Overview and comparison of AI strategies of European Member States*. Paper presented at the Electronic Government, E-Democracy and Open Government, and Electronic Participation: Joint Proceedings of Ongoing Research and Projects of IFIP WG 8.5 EGOV-CeDEM-ePart 2020, Linköping, Sweden.
Vassilakopoulou, P., Grisot, M., & Aanestad, M. (2017). Friction forces and patient-centredness: Understanding how established logics endure during infrastructure transformation. *Health Informatics Journal, 25*(2), 361–371. https://doi.org/10.1177/1460458217712053
Wagenaar, R. W. (2006). *Governance of shared service centers in public administration: Dilemmas and trade-offs*. Paper presented at The 8th international conference on Electronic commerce (ICEC 2006), Fredericton, New Brunswick, Canada.
Wendt, A. E. (1987). The agent-structure problem in international relations theory. *International Organization, 41*(3), 335–370.
Whittlestone, J., Nyrup, R., Alexandrova, A., Dihal, K., & Cave, S. (2019). *Ethical and societal implications of algorithms, data, and artificial intelligence: A roadmap for research*. Nuffield Foundation.
Winner, L. (1977). *Autonomous technology: Technics-out-of-control as a theme in political thought*. MIT Press.
Winner, L. (1980). Do artifacts have politics? *Daedalus, 1*. Retrieved from https://login.e.bibl.liu.se/login?url=http://search.ebscohost.com/login.aspx?direct=true&db=edsjag&AN=jstor.10.2307.20024652&site=eds-live

Actors and Intentions in the Dissemination of Robotic Process Automation in Social Work

Agneta Ranerup and Lupita Svensson

1 Introduction

Various types of automated decision-making are discussed, promoted (cf. SALAR, 2018a; Spielcamp, 2019) and disseminated in numerous areas of the public and private sectors. Moreover, robotic process automation (RPA), which has been described as 'AI-Lite' because it cannot develop algorithms but is directed by software programming (Wirtz et al., 2019), is increasingly being implemented in public and private sector contexts (cf. Asatiani et al., 2019; Houy et al., 2019). The RPA tool, or the software 'robot', is equipped with the capacity to analyse structured data based on algorithms and, therefore, can perform simple, repetitive tasks. Based on this analysis, the RPA can suggest an outcome of such a process, or what can be characterised as 'a decision' (Houy et al., 2019). Since 2017, RPA has been used to make decisions regarding the allocation of social assistance (e.g. economic support for food and housing) within social service organisations in a growing number of municipalities in Sweden (Ranerup & Henriksen, 2019, 2020; Svensson, 2019b). Social assistance is a form of economic support for people in need that is approved by social service caseworkers based on applications from clients. The Nordic countries have developed a similar welfare structure through which social assistance must be the last resort and is often provided at a minimum level. (For a recent example from Norway, see Gjersøe [2021] and NAV [2021].) The use of RPA in social work has been controversial, triggering a societal debate concerning case-workers' discretion (Persson, 2018) and the legal status of the algorithm used in

A. Ranerup (✉)
University of Gothenburg, Gothenburg, Sweden
e-mail: agneta.ranerup@ait.gu.se

L. Svensson
University of Lund, Lund, Sweden
e-mail: lupita.svensson@soch.lu.se

G. Juell-Skielse et al. (eds.), *Service Automation in the Public Sector*, Progress in IS,
https://doi.org/10.1007/978-3-030-92644-1_7

'decision-making' (Erkers & Vinge, 2020; Kaun, 2021) in social work in contrast with what is used in other areas, like state administration (Wihlborg et al., 2016). This chapter provides an overarching view of important actors and intentions and their activities to influence the dissemination and implementation of RPA in social work based on experiences from Sweden.

Information technology (IT) has long been used in social work. Laurent (2008) described the introduction of IT into Belgian social work, proposing the influence of two logics in this field: 'the logic of computerisation' and the 'logic of social work'. Actors using IT in social work may have different intentions or aims. For example, Devlieghere et al. (2017) interviewed directors, policy advisors and staff members who spoke of administrative, policy, care and economic reasons for implementing IT. At the same time, based on interviews with managers and other actors, Gillingham (2018a, b) encouraged the development of a critical attitude with the intention to use IT to support organisational change, not to lead it; the studies described the risk of transforming a bureaucracy with its positive aspects to an outright technocracy.

Further, IT may also feature in decision-making processes in social work. Artificial intelligence—using predictive algorithms in the analyses of large volumes of administrative data—has recently been described as a promising means for supporting decision-making regarding issues related to children and families (Gillingham, 2021). Another more direct use is related to applications for social assistance that may be partially or wholly automated using technology, such as RPA. Researchers have raised the concern that caseworkers' attention to clients' individual needs may be minimised in decision-making related to social assistance when RPA is used (Gustafsson & Wihlborg, 2019). Petersen et al. (2020) described caseworkers' use of discretion or independent right to decide based on available laws (Lipsky, 2010) as cooperative endeavours involving colleagues and clients that are reliant on individual consultative skills. However, in automated decision-making, a certain degree of uniformity and simplification is necessary (Petersen et al., 2020). From a broader perspective, Ranerup and Henriksen (2019) analysed the 'value positions' aspired to that underlie the implementation of RPA in social assistance. Here, improved efficiency and service in the form of high-quality contact between caseworkers and clients were prioritised. Further, Ranerup and Henriksen (2020) more closely analysed the distribution of discretion between humans and technology. They found a repertoire of humans and technologies involved in actual decision-making processes instead of a single caseworker and the RPA. In sum, social work is deeply associated with meetings between clients and caseworkers (Laurent, 2008), caseworkers' discretion (Lipsky, 2010; Petersen et al., 2020; Ranerup & Henriksen, 2020) and the individual needs of clients (Gustafsson & Wihlborg, 2019).

In an overview of research regarding IT in social work, Lagsten and Andersson (2018) found interesting areas in need of further research: IT governance was one of the most under-researched topics. Involving researchers and theories from the information systems discipline in research on IT in social work with the intention of understanding the groups of human actors involved and the technological artefacts that may themselves be considered important actors was considered crucial (Callon, 1986; Latour, 2005). In contrast to studies of involved actors' views of technology implementation in social work, such as those described by Devlieghere et al. (2017) and Gillingham (2018a, b) and case studies on the use of RPA in social work (Gustafsson & Wihlborg, 2019; Petersen et al., 2020; Ranerup & Henriksen, 2019), the present study provides an overarching picture of actors that participate in the debate about and dissemination of RPA in the public sector area of social work in Sweden. Important focus is placed on their intentions or aims communicated in association with the dissemination of RPA. We also provide a brief overview of the actual implementation of RPA. In line with Lagsten and Andersson (2018) and suggestions from the field of social work (Ballantyne, 2015), our analytical approach was inspired by the actor-network theory (ANT) that problematises the relationship between humans and technology (Callon, 1986; Latour, 2005). Equally important, our research provides insights about the dissemination of RPA with relevance from a general public sector perspective.

The research questions for this study are as follows:

- What actors, intentions and network formations appear in the dissemination of RPA in the management of applications for social assistance in Sweden?
- What are the key, sometimes conflicting, intentions from a social work and a general public sector perspective?

In the next section, our methodology and data are described. This is followed by an empirical account of the dissemination of RPA in the management of applications for social assistance in which important events, participating actors and their intentions are described. Lastly, a discussion and several conclusions are presented.

2 Methodology

The methodology for this study was based on providing a qualitative, event- or phase-based account and analysis (Cho et al., 2008; Ranerup, 2012) of important involved actors and their intentions related to the use of RPA in social work. The events or phases serve as a simple means for structuring the otherwise extensive account of important activities performed by the actors involved. Inspired by the most basic aspects of the theoretical perspective of ANT (Callon, 1986; Latour, 2005), we attempted to 'follow the actors', their respective and sometimes conflicting intentions and their formation of relationships or networking in the process of promoting, disseminating and implementing RPA in the management of applications for social assistance in Sweden. These actors were representatives from various organisations (national, municipal and other), and they had certain aims and intentions that they wished to 'inscribe' or implement into the arrangements for disseminating RPA. The types of technologies that are mentioned herein are, themselves, also perceived as important actors (Callon, 1986; Latour, 2005). Our choice of an event- or phase-based account and analysis (Cho et al., 2008) was

Table 1 Sources for empirical data

Event	Data source	Role in organisation	Role in the process	Number of instances
The Trelleborg Case	Interviews and oral presentations	Managers, politicians and caseworkers	Drivers and implementers of change, case management related to social assistance	21
	Documents	Internal and external documents regarding the change process	Communication of official policies, assessments and change promotions	43
The SALAR Case	Interviews and oral presentations	Project leaders and consultants from SALAR	Drivers and implementers of change	12
	Documents	Plans and descriptions from the dissemination project	Description of intentions	6
National survey of RPA implementation	Survey	Social services in Swedish municipalities	Description of implementation of RPA	290
	Interviews	Caseworkers, managers, politicians	Drivers and implementers of change	26
	Documents	Plans and descriptions from the dissemination projects	Description of intentions and process	10
National news with relevance to RPA	Articles in national press and TV	News and debates	Description of controversies regarding new laws and algorithms	7

prompted by our objective to cover a process with a spectrum of related actors that were involved in the dissemination of RPA, but not in the form of a more 'traditional' translation process, beginning with an initial network formation and a clear goal (cf. Callon, 1986). Furthermore, our approach is in line with Ballantyne's (2015) suggestion to use ANT in research on human services.

The data were collected between September 2017 and March 2021 (see Table 1). Our data collection consisted of four central components: (1) 'The Trelleborg Case', specifically, the municipality of Trelleborg and its implementation of RPA in 2017, along with its prehistory; (2) 'The SALAR Case', specifically, the national Swedish Association for Local Authorities and Regions (SALAR) and its dissemination activities in 2018–2020; (3) a national survey on RPA implementation in social work in Swedish municipalities created by one of the authors in 2019; and (4) data from appearances in the national news related to the use of RPA in social work during this period, including suggestions for changes in the law regulating RPA in the public sector in March 2021.

The account was made by 'following the actors' that could be considered important and relevant in a fair description of activities related to RPA in the case management of applications for social assistance. The two co-authors have been involved in research and praxis in the field of RPA in social work since 2017. In line with studies such as Cordella and Hesse (2015) and Ranerup (2012), Sect. 3 provides a broad event-based account and Sects. 4 and 5 more directly apply a few concepts from ANT (cf. actors, intentions and networking with the aim of enrolment). The account is a joint product that was discussed in a seminar with researchers in the area in November 2020. The concluding discussion compares the various actors and intentions to capture what the dissemination involves from the perspective of social work and from a more general public sector perspective.

3 An Empirical Account of the Dissemination of RPA in Applications for Social Assistance

3.1 An Innovative Municipality's Early Implementation of RPA

Since the beginning of the 2010s, the municipality of Trelleborg has worked to change the focus of the management of applications for social assistance. The intention was to gradually transition from providing social assistance into the more accepted position of helping applicants to become more self-sufficient (Trelleborg, 2015). As part of this revised focus, individuals who apply for social assistance are given the opportunity to participate in education and obtain practical experience or a job. Those with more serious problems are offered special support. The municipality was active in promoting its new model for providing social assistance in the national political arena. It published descriptions of what changes were made (Trelleborg, 2015), and numerous visitors were welcomed. To enable these changes, the social assistance application process was streamlined to make more time available for caseworkers to meet with applicants. After the process was streamlined an e-application was implemented as of September 2015.

In 2016, Trelleborg began to use RPA in a few municipal administration areas. In 2017, it was introduced into the management of applications for social assistance (Trelleborg, 2017a). A politician provided the following reflections on the implementation of e-applications and the RPA:

> It is natural to make things more efficient. [...] I don't understand why we don't do this even more. This is the direction our society is taking (Politician 1, Labour Market Board, Trelleborg, 25 September 2017).

The RPA was designed so that caseworkers' participation would secure the quality of the work it performed. The RPA made the full decision in approximately 40% of the less complex cases pertaining to applications for social assistance and partially contributed to decisions for cases with a higher degree of complexity (Trelleborg, 2017a). The consultants involved emphasised the importance of preserving the competence of professional caseworkers when RPA is implemented

(Oral presentation for municipalities about RPA in Trelleborg, Valcon representative, 6 October 2017). A central part of the new management model was the service provided through regular meetings between caseworkers and applicants:

> But as I see it, the parts that are most important are [not the RPA but] the support we offer; we provide a high-profile service (Manager 1, Labour Market Agency, Trelleborg, 17 October 2018).

As part of these efforts, the municipality of Trelleborg applied to SALAR for funding for a project that aimed to test a model for disseminating innovative forms of digitalisation that represented a new way of working by focusing on promoting self-sufficiency, which became known as 'the Trelleborg model'. During the spring of 2017, the project began with 12 municipalities participating in 4 on-site meetings. The digitalisation and the RPA were mentioned a few times in a newsletter, for example, when a few caseworkers described their experiences:

> It is not about the robot replacing the human—quite to the contrary. [. . .] It makes our work easier and enables us to provide a better service in the part of the process where the focus is on becoming more self-supporting (Trelleborg, 2017b).

An interviewee from one of the 12 participating municipalities summarised the intentions of the municipality when referring to an effect of participating in the dissemination project: 'Yes, there was much talk about robotisation in that project' (Manager 2, Social & Labour Market Agency, Mölndal, 2 May 2019).

3.2 The Early Promotion of Robotic Process Automation (RPA) by the Swedish Association for Local Authorities and Regions (SALAR)

All 290 municipalities of Sweden are members of SALAR, the national organisation for local authorities. SALAR published reports highlighting that the implementation of automated processes in municipal administration operations saves time that can be spent on other activities and praising automation for also promoting. . .

> . . .higher quality and preserved rule by law in decisions from the perspective of citizens and companies. Another motive [for implementing automation] might be to improve the work environment for the co-workers [in municipal administration]. Still other motives can be improved efficiency and effectiveness. [. . .] However, lower costs are not necessarily an effect, but sometimes [automation results in] shorter waiting times for the individual in case management and [in] a process with less complex decision processes without inbuilt flaws (SALAR, 2018b, p. 10).

The more general reason for introducing automation into the public sector was the future demographic situation, which led to difficulties in recruiting for and financing the public sector (SALAR, 2018a, p. 2). The report also provided advice for managing processes that involve the introduction of RPA. SALAR assigned importance to formulating the reason behind, or 'the why' justifying, the changes related to the implementation of RPA.

A variety of opinions were expressed about the project run by Trelleborg from 2017 to spring 2018. According to one SALAR representative, the main aspect of the Trelleborg project activities was not the focus on labour market issues but 'the automation of a process [performed] where too many people work' (Manager working with digitalisation, SALAR, 4 May 2018), a feature that attracted interest among other municipalities in Sweden as well. On the other hand, other informants claimed that Trelleborg's dissemination project did not focus on automation but on 'moving from social work to a labour market perspective' (cf. Consultant 1 appointed by SALAR, 26 September 2019).

During 2017–2018, interest in RPA increased among Swedish municipalities. Therefore, SALAR initiated a national project to stimulate the dissemination and exchange of accounts of RPA experiences. The project was scheduled for implementation from spring 2019 to August 2020. The chief information officer (CIO) of SALAR explained its larger aim as follows:

> SALAR now takes this initiative to be part of necessary changes. The competence that professionals possess can in the future be used to strengthen individuals' connections to the labour market [...] instead of [for] calculating the exact benefit payments (CIO of SALAR, 1 February 2019).

SALAR sponsored a webinar on 7 November 2018 that outlined the details of the dissemination project, which was presented by consultants appointed from PwC who previously led the implementation of RPA in Trelleborg. The theme of the webinar was 'automation in social assistance decisions'.

3.3 *Critical Debate in Society*

During the period 2018–2019, many people were interested in implementing RPA for decision-making in social work, but some still viewed such a move with mixed feelings. A representative from one of the 12 municipalities who participated in the dissemination project run by Trelleborg (see Sect. 3.1) indicated that they would implement the new focus on labour market issues and digitalise the management of social assistance. The role of RPA in the management of applications for social assistance was also addressed: 'But it is still a bit difficult to understand that it is a "digital co-worker" that you will have as a colleague' (Manager 1, Social & Labour Market Agency, Mölndal, 8 March 2019).

In conjunction with the dissemination project run by Trelleborg, a participating municipality (Kungsbacka) was the subject of a national news story because caseworkers had left the municipality due to the new focus on self-sufficiency and the introduction of 'a robot'. Moreover, the dissemination of the Trelleborg model of working that focused on labour market issues and self-sufficiency prompted criticism for 'too much copying' of the method (cf. Persson, 2018) and for issues regarding the implementation of RPA associated with it. Furthermore, although representatives of the Union for Professionals (Akademikerförbundet SSR) Union and others were active in the debate about the status of the 'algorithms' used in decision-making (see Kaun, 2021), union members still voiced concerns regarding the use of automation in the management of social assistance and the impact RPA would have on social work (Svensson, 2019b). Therefore, the Union decided at its congress in 2018 to call for a scientific study of the short-term and long-term effects of the 'Trelleborg model' for social assistance.

3.4 Actual Implementation of RPA in Swedish Municipalities

In Sweden, municipal self-government prevails; indeed, the country is home to 290 municipalities with rather varied areas and populations. The inventory check in the scientific study conducted in 2019 revealed that just under 30% (or 86) of the 290 municipalities linked a digital e-service to the application for social assistance (source in this section: Svensson, 2019a, b), yet only 16 municipalities had implemented some form of a digital automation function for social assistance case management. One of the biggest challenges identified was connecting the e-service to the municipality's administrative platforms in the case management systems to then be able to introduce an RPA (often called a 'robot'). Numerous municipalities were linked to these systems through long contract periods. However, only a few technical companies who dominated the market were responsible for the administrative platforms in the case management systems.

The intentions underlying the introduction of RPA were to free up some of the time caseworkers spent on administrative work so they could perform qualified social work instead. Several municipalities emphasised the automated decision-making itself was not driving the change in systems; rather, the transition was centred on enhancing organisational development to better meet clients' needs. Of the 16 municipalities that had implemented RPA, only one (Trelleborg) allowed the 'robot' to make decisions, whereas the remaining municipalities used it merely to support decision-making.

The results indicated that implementing automated decision-making or RPA is a slow process. One of the success factors that emerged was clearly set goals (answering 'why'), preferably anchored in politics. Developing an RPA is an on-going task, something that the social service administrations had not considered. Instead, any problems that were encountered were perceived as indicating a 'fault' in the algorithms and considered evidence that the programme did not work. A crucial and problematic aspect that numerous interviewees recalled was the technical solution of an RPA and its integration with the case management system.

In the municipalities where the automation and the automated decision-making system worked and, therefore, provided caseworkers with additional time for other activities and qualified social work, questions such as 'what is qualified social work' and 'will our services be withdrawn now' arose. The interviewees described those who worked with the administration of the applications as becoming even better administrators, while on the other hand, social work did not represent key knowledge or skills in the labour market units in which some clients were receiving assistance to become self-supporting.

3.5 *The National Dissemination Project Run by SALAR*

The national dissemination project SALAR initiated in 2019 ran through spring 2020 and included four phases or meetings according to the method designed by the experienced consultants appointed, who had rich experiences from holding leading positions in Trelleborg. The four phases were as follows: (1) begin by introducing RPA and planning, (2) change management, (3) systematic follow-up on activities and (4) finalisation and measurement of results. Between these meetings, the consultants were available as advisors (SALAR, 2018b, p. 9). The change management activities resulted in 'enabling the partaking municipalities to implement a "digital co-worker" in decisions about social assistance' (SALAR & PwC, 2018, p. 11). In spring 2019, 12 municipalities were enrolled in the first round for the national project, and another group of 12 municipalities was enrolled in September 2019. Each round ran for approximately one year.

Formulating an answer for 'why' RPA was being implemented was important, but the answers varied across the municipalities involved. The consultants described this viewpoint, as well as the difficulties involved in formulating the 'why':

> A majority of the municipalities involved in the first round want to implement a 'digital co-worker' in order to support more citizens to become self-supported. [...] Then there are variations that in different ways are related to allocating resources to activities that create value for the citizens. [...] [But the issue of 'why'] must be returned to later since this is not as simple as they initially expected (Consultant 1, PwC, 26 September 2019).

The project leaders from SALAR continued:

> Many [people] ask if the intention is to implement the Trelleborg model. But it isn't (Project leader 1, SALAR, 16 September 2019).

The importance of implementing an e-application for use by many applicants and the complicated role of the case management systems in the municipalities were also mentioned. Across the country, according to the interviewed project leader (25 May 2020), 80–90 municipalities decided to initiate this kind of process. The most important intentions among municipalities were about developing their case management process, deciding the rules in automated decision-making that the RPA must follow and the extent to which RPA should be involved in the case management process (Consultant 2, PwC, 11 March 2021).

3.6 Increasing Experiences and Potential New Laws

The number of municipalities with comparatively lengthy (2.5–3.5 years) experiences with using RPA for social assistance decision-making has increased; Trelleborg, Nacka, Strängnäs and Norrtälje are the most experienced. The technology is further developed in the municipality after it is implemented. For example, in Trelleborg, since the summer of 2019, digital support has been available for submitting denial of claims decisions. In contrast to when the e-application was introduced, paper documents must now be submitted in a few situations, for example, when entering a contract to rent an apartment. By June 2020, the RPA could support making a few types of negative decisions. In 2020, over 85% of social assistance applicants in Trelleborg used the e-application (Manager 2, Labour Market Agency, Trelleborg, 25 June 2020).

A significant potential for the dissemination of RPA in local government administration at large is said to exist (Houy et al. 2019). Indeed, the dissemination process has gained momentum. However, the consultants involved in the national project run by SALAR during the period spring 2019–August 2020 commented on existing complexities in implementing RPA in social work:

> [Making decisions about social assistance] is not the easiest area [in the municipal administration] to begin with. It's not uncontroversial either (Consultant 2, PwC, 26 September 2019).

The future legal status of automated decisions is part of an animated debate, as are suggestions for a new legal framing. A proposal for a draft of a new legal text was presented 24 March 2021 (The Government, 2021). At the same time, the union SSR continued to participate in the national debate about the lack of transparency of algorithms and its lawfulness when used like it was used in Trelleborg (cf. Erkers & Vinge, 2021).

4 Case Analysis and Discussion

4.1 Actors and Relations in the Dissemination of RPA in Social Work

Our event-based account (Cho et al., 2008) of actors and intentions (Ballantyne, 2015; Callon 1986; Latour, 2005) includes various actors representing multiple types and levels of organisations (Table 2, left column). The first is SALAR, the national organisation that tried in many ways (reports, dissemination projects, seminars) to build active relationships between the RPA technology and interested municipalities. This can be considered a 'soft' governance initiative (Lindgren et al., 2021) for RPA dissemination. Another national actor is the union SSR that, against the background of a more critical standpoint, took part in the debate about RPA and

Table 2 A summary of actors and intentions

Actor	Intentions and related activities
SALAR	Meeting municipalities' interest in RPA, saving resources and time spent on administration, counteracting difficulties to recruit personnel
Union SSR	Criticising the new focus in social work, questioning the legality of decisions and initiating research about RPA
Trelleborg	Promoting a new model for helping applicants to become self-supporting, implementing e-applications and RPA, disseminating the new model and the RPA
Municipalities implementing RPA	Expressing an interest in the new model with a focus on self-support and the RPA, formulating intentions with the RPA
Consultants	Helping with many aspects of dissemination projects, especially formulating a 'why', promoting its services
Clients	Being willing to submit e-applications, receive decisions from the RPA
Technologies	Providing 'digital material' through e-applications and necessary infrastructure for the RPA, thereby suggesting or making 'decisions'

its consequences and brought forward new knowledge about its actual use. Interestingly, this was partly done in reaction to the activities of a municipal actor, Trelleborg, and its active dissemination of the model for social assistance with a focus on RPA and self-support. Equally important, other municipal actors took part in the municipality of Trelleborg's and SALAR's dissemination activities or, on their own initiative, implemented RPA in social work (Svensson, 2019a, b).

An additional actor was the project technical consultants. In our case, with a focus on early dissemination of RPA in the relatively complex area of social work, they had been leading actors in Trelleborg. We also noticed that, as a formal part of preparing SALAR's dissemination project, a broader search for consultants was conducted. However, the qualified consultants previously connected to the innovative Trelleborg case were ultimately selected. Other actors, such as companies like PwC, Valcon and similar businesses, are needed to create a successful relationship between municipality, RPA, and other components of the necessary technological infrastructure, as well as to formulate the intention (the 'why') behind the implementation. Our study illustrated that companies like these are actively using experiences from innovative projects with RPA to promote their services as they try to enrol municipalities as customers.

Other important actors are, of course, the clients who apply for social assistance. They appeared in all municipalities but especially in conjunction with the introduction of e-applications, as well as in the new forms of work characterised by more meetings between caseworkers and clients. The intention of providing good service to applicants, albeit with various interpretations of what constitutes good service, was considered important by most involved actors. However, a basic condition for the RPA is that applicants must use the e-application. For this to happen, a significant degree of trust in applicants must exist so that the paper components of the application process can be changed and support can be provided to applicants (Ranerup &

Henriksen, 2020). As noted, in Trelleborg recently, approximately 85% of the social assistance applications were submitted via the e-application.

Obviously, technological actors are also involved. An e-application must be implemented in conjunction with or before initiating the RPA mechanism, which must be used by a large proportion of applicants. Equally important, the RPA itself is part of the process. Its implementation must be a result of public procurement and design processes in which local representatives of social workers participate to form the decision rules or algorithms embedded in the RPA.

Other technological actors of note are the case management systems. These systems have appeared in the national survey (Svensson, 2019b) and in the account of SALAR's national dissemination project activities. For various reasons, the capacity of this type of technology to integrate the e-application and the RPA into its platform is problematic. Thus, the case management systems will, at least partially, in a more persistent manner, guide the development of automation. This implies a shift in what governs this development because the handling of social assistance, or the 'core process' and the 'support processes', like in the administrative systems, has become equally important. Consequently, the competence of and relationship with human actors in the form of technical consultants is vital.

In sum, network formation in the dissemination of RPA can include national actors that actively try to enrol interested municipalities through various activities, but in our study this was also accomplished through appointed consultants in a high-profile innovative case (Trelleborg). Also, 'critical' forces or actors like unions (e.g. SSR) may be involved that also actively involved the Trelleborg case in their discussions on algorithms, transparency and lawfulness, as in the case studied here. Thus, an important element of the dissemination of RPA can be high-profile cases that are part of the 'travel of ideas' (Czarniawska & Joerges, 1996).

Network formation at the local level (Lindgren et al., 2021; Ranerup & Henriksen, 2020) naturally must involve actors like caseworkers, clients and local staff with knowledge about various forms of technology. However, more specific details are outside the scope of the present study.

4.2 Intentions for the Dissemination of RPA from a General and a Social Work Perspective

In our account, summarised in Table 2, actors at different levels promoted intentions that they actively tried to inscribe into the process and its outcome in terms of a new type of case management for social assistance with RPA in place.

We also identified more general intentions put forward by many actors related to the dissemination of RPA for streamlining application processes, introducing e-applications and improving efficiency and effectiveness. Other intentions included satisfying municipalities' interest in RPA and their anticipated future difficulties with recruiting personnel for the municipal administration. Just as important was the intention we uncovered to formulate a 'why' behind the implementation. All these intentions provide, we argue, a simple but relevant summary of the changes that must be incorporated into, or become a part of, future public administration processes with RPA. In other words, these intentions were relevant for many areas of municipal administration in which RPA is implemented (Houy et al., 2019; Lindgren et al., 2021) in addition to social work.

Another type of intention was related to questioning the legality of using RPA in decision-making and the transparency of algorithms used as part of those decision-making processes. These were mostly, in our case, put forward by the union SSR. However, in this case, actors representing many municipalities except one have taken similarly critical positions against RPA as a formal decision maker: the Trelleborg municipality actually defined the outcome of the RPA operation as a formal 'decision' (Ranerup & Henriksen, 2020). We contend that these intentions were also of a general character and, therefore, of relevance to many areas of the public sector in which RPA is not allowed or is being questioned.

Finally, in our account, we detected intentions presented by actors related to changing the focus of social assistance case management to strengthen the capacity to support applicants in their efforts to become self-sufficient. The issue of formulating a 'why' behind the implementation of RPA is also related to this intention, as well as to the more general situation of implementing RPA in areas other than social work. Intentions related to improving efficiency and effectiveness through RPA can contribute to the introduction of new activities, such as those made possible by caseworkers having more time to spend with clients in social work, as well as in other areas in the public sector (SALAR, 2018b).

The intentions put forward by actors are rather 'natural' and unsurprising (promoting e-applications, efficiency and effectiveness, legality). However, conflicts can also exist between actors and their intentions about, for example, the more general aspects of legality and transparency. In the area of social work, conflicts can also relate to the larger programme behind the changes implemented through RPA, like launching new models behind activities. Possible examples in the area of social work include intentions and activities to enhance individuals' ability to become self-supporting and a labour market perspective (Gjersøe, 2021; NAV, 2021). Thus, through the context of our study, we see that intentions for the dissemination of RPA may serve to question the core of social work and what goals it should strive to meet. This includes the more detailed organisation of the case management process, as well as the competence and role of different types of professional caseworkers within a new model with different aims. In our study this aspect was, perhaps not surprisingly, emphasised by unions representing the professional caseworkers.

Interestingly, our case showed that a national actor like SALAR emphasised more general intentions about RPA behind its dissemination project rather than changing the focus of social assistance in line with a labour market perspective and the 'Trelleborg model'. This can be seen as a way of drawing a clear line between the organisation's activities and a controversial model or case (Erkers & Vinge, 2020, 2021; Kaun, 2021; Persson, 2018) and is in line with a more general interest in disseminating RPA (SALAR, 2018a).

5 Conclusions, Limitations, and Further Research

A conclusion in our study is that the dissemination of RPA in the public sector area of social work may be inhabited by actors at the national, municipal and local levels. They are part of the dissemination process in that they actively try to enrol other actors to implement RPA or, alternatively, to question aspects of its use in social work.

A further conclusion is that an innovative case of RPA implementation can serve as an initiator, a source of experiences and a critique beyond its origins that lie in a municipality with 45,000 inhabitants (Trelleborg). A national organisation like SALAR may actively, albeit somewhat indirectly, use it as a part of 'soft power' to disseminate RPA (Lindgren et al., 2021). This may happen in other areas as well, but may have gained a specific power due to the area of social work and the controversies surrounding it (Kaun, 2021). A long tradition has been made of distinguishing between 'the logic of computerisation' and the 'logic of social work' (Laurent, 2008). Our results also show that actors like unions (the SSR in this study) may strive to exercise a kind of 'discursive power' regarding laws and issues about transparency but that they actively used the specific example of Trelleborg in this case.

In addition, we saw in our study that RPA became an important aspect in projects aiming to test new ways of disseminating innovative forms of digitalisation, although this was not true in the beginning (see Sect. 3.1). In these and other ways, we claim that RPA has served as an actor in its own right. However, whether RPA receives a far too dominant position as a source of automation and digital innovation in public sector contexts is debatable since it is not without problems (Lindgren et al., 2021) and because more advanced technologies exist (Houy et al., 2019). Also, technological actors include not only the RPA but also the e-application and the case management system. Thus, in the dissemination of RPA in social work, technological actors need to be taken seriously and not only seen as a source of 'technomagic' in the form of excessive interest or as the source of power for certain actors (Gillingham, 2018b).

We have captured an account of intentions put forward by actors in social work (Devlieghere et al., 2017; Gillingham, 2018a, b) in relation to RPA. We conclude that most of the intentions put forward are relevant in many areas of public sector (Houy et al., 2019) in relation to RPA (e.g. efficiency, effectiveness, legality and transparency). In contrast, our specific area of study showed that RPA may be associated with changes that question fundamental issues related to implementation, expressed in terms of the 'core values' of social work in relation to a labour market perspective (Gjersøe, 2021).

Our conclusions contribute to research regarding circumstances and trends in relations on the dissemination of RPA in the public sector (Houy et al., 2019), particularly considering the sensitive area of social work (Petersen et al., 2020; Ranerup & Henriksen, 2019; Svensson, 2019a). Our theoretical framework has the intention of strengthening the understanding of technology in social work

(Ballantyne, 2015; Lagsten & Andersson, 2018) as a public sector area (Cordella & Hesse, 2015; Houy et al., 2019). The actor-intention perspective emphasises the range of aspects that may lie behind RPA. These intentions must, at least ideally, be formulated against the fundamental aspect that RPA is not, as mentioned, a value in itself but, instead, must meet its most fundamental role of fulfilling the aims of a welfare state and its citizens. Thus, it is important to apply a critical perspective on issues related to design and access (Coles-Kemp et al., 2020).

A limitation of our account is its overarching level of analysis. Therefore, further research must involve much more detailed studies of the situation of caseworkers and their discretion from a longer time perspective in view of their new competence and changed experiences. In addition, the potential for programming the RPA for different types of more advanced decisions, including negative ones, in contrast to the discretion of caseworkers in making and communicating decisions must also be considered (Gustafsson & Wihlborg, 2019; Petersen et al., 2020; Ranerup & Henriksen, 2020). The entire range of technologies (e-application, case management system, RPA) and their enlarged roles may also be important in influencing the discretion or capacity of caseworkers to decide.

In Sweden, a proposal for a draft of a new legal text regarding automated decision-making in the public sector was presented on 24 March 2021 (The Government, 2021). According to the present Public Administration Act (2017: 900), decisions can be made by one executive civil servant alone, by several jointly or *automatically,* which means that there is *no* requirement for human participation. The Public Administration Act generally applies in the state and municipal administrations but is subordinated to deviating provisions in other regulations, for example, the Local Government Act (2017: 725). The Local Government Act is the main regulation for the municipalities and is based on the principle that decision-making is based on a local politically appointed board related to, for example, social services. The act allows for the possibility of delegating decisions but always includes a human decision maker. In the new suggested law, the government considers that municipalities as well as the state must be able to use automated decision-making to take advantage of digitalisation in administration. One issue treated in the public investigation and the suggested legal texts is how to harmonise the two regulations. Another issue is the principle of openness and the regulation of documentation and transparency as part of the Local Government Act (2017: 725). Therefore, the new situation with an expanded use of technology, on the one hand, and, on the other, the relationship between the described laws and perspectives in the public sector area of social assistance present an additional challenge worthy of investigation.

Funding Thanks are due to Swedish Research Council for Health, Working Life and Welfare grant no. 2019-00710 for funding our research.

References

Asatiani, A., Penttinen, E., Rinta-Kahila, T., & Salovaara, A. (2019). Implementation of automation as distributed cognition in knowledge work organizations: Six recommendations for managers. In *Proceedings of international conference on information systems*. Association for Information Systems.

Ballantyne, N. (2015). Human service technology and the theory of the Actor-Network. *Journal of Technology in Human Services, 33*(1), 104–117.

Callon, M. (1986). Some elements of a sociology of translation: Domestication of the scallops and the fishermen in St Brieuc Bay. In J. Law (Ed.), *Power, action and belief: A new sociology of knowledge* (pp. 196–233). Routledge.

Cho, S., Mathiassen, L., & Nilsson, A. (2008). Contextual dynamics during health information systems implementation: An event-based actor-network approach. *European Journal of Information System, 17*, 614–630.

Coles-Kemp, L., Ashenden, D., Morris, A., & Yuille, J. (2020). Digital welfare: Designing for more nuanced forms of access. *Policy Design and Practice, 3*(2), 177–188.

Cordella, A., & Hesse, J. (2015). E-government in the making: An actor network perspective. *Transforming Government: People, Process and Policy, 9*(1), 104–125.

Czarniawska, B., & Joerges, B. (1996). Travel of ideas. In B. Czarniawska & G. Sevon (Eds.), *Translating organizational change* (pp. 13–47). de Gruyter.

Devlieghere, J., Bradt, L., & Roose, R. (2017). Policy rationales for electronic information systems: An area of ambiguity. *British Journal of Social Work, 47*, 1500–1516.

Erkers, H., & Vinge, S. (2020). Unauthorized algorithm make decisions in social work [Obehörig algoritm fattar beslut i socialtjänsten]. *Svenska Dagbladet, 18*. January 2020. Akademikerförbundet SSR.

Erkers, H., & Vinge, S. (2021). There must be full transparency regarding algorithms in the public sector [Full insyn måste råda i offentliga algoritmer]. *Dagens Nyheter, 26 July*, 5.

Gillingham, P. (2018a). Decision-making about the adoption of information technology in social welfare agencies: Some key considerations. *European Journal of Social Work, 21*(4), 521–529.

Gillingham, P. (2018b). From bureaucracy to technocracy in social welfare agency: A cautionary tale. *Asia Pacific Journal of Social Work and development*. https://doi.org/10.1080/02185385.2018.1523023

Gillingham, P. (2021). Algorithmically based decision support tools: Skeptical thinking about the inclusion of previous involvement. *Practice, 33*(1), 37–50.

Gjersøe, H. M. (2021). Providing help or restrictions? Frontline workers´ understanding of behavioral conditionality for health-related social insurance and social assistance benefits in a joined-up governance context. *European Journal of Social Work*. https://doi.org/10.1080/13691457.2021.1934412

Gustafsson, M. S., & Wihlborg, E. (2019). "It is always an individual assessment": A case study on challenges of automation of income support services. In Lindgren I et al. (Eds.) *EGOV 2019, LNCS* (pp. 45–56). Springer.

Houy, C., Hamberg, M., & Fettke, P. (2019). Robotic process automation in public administration, Digitalisierung von Staat und Verwaltung. *Lecture Notes in Informatics (LNI)* (pp. 62–74).

Kaun, A. (2021). Suing the algorithm: The mundanization of automated decision-making in public services through litigation. *Information, Communication & Society*. https://doi.org/10.1080/1369118X.2021.1924827

Latour, B. (2005). *Reassembling the social: An introduction to actor network-theory*. Oxford University Press.

Lagsten, J., & Andersson, A. (2018). Use of information systems in social work – Challenges and an agenda for future research. *European Journal of Social Work, 21*(6), 850–862.

Laurent, V. (2008). ICT and social work: A question of identities? In S. Fischer-Hübner, P. Duquenoy, A. Zuccato, & L. Martucci (Eds.), *The future of identity in the information* (pp. 375–386). Springer.

Lindgren I, Toll D, Melin U (2021) Automation as a driver of digital transformation in local government, DG.O'21, June 09-1, ACM.
Lipsky, M. (2010). *Street-level bureacracy: Dilemmas of the individual in public services* (30th anniversary ed.). Russel Sage Foundation.
NAV. (2021). *Economic assistance [Økonomisk sosialhjelp].* Norwegian Welfare and Labour Organization. https://www.nav.no/sosialhjelp/nodsituasjon. Accessed 25 May 2021.
Persson, E. (2018). The successful Trelleborg-model causes resistance: "Too much copying" [Succémodellen i Trelleborg möter motstånd: "Man kopierar"]. *Arbetsvärlden.* January 12. https://www.arbetsvarlden.se/succemodellen-i-trelleborg-moter-motstand-mankopierar/. Accessed 30 September 2018.
Petersen, A. C. M., Christensen, L. R., & Hildebrandt, T. T. (2020). The role of discretion in the age of automation. *Computer Supported Cooperative Work.* https://doi.org/10.1007/s10606-020-09371-3
Ranerup, A. (2012). The socio-material pragmatics of e-governance mobilization. *Government Information Quarterly, 29,* 413–423.
Ranerup, A., & Henriksen, H. Z. (2019). Value positions viewed through the lens of automated decision-making: The case of social services. *Government Information Quarterly, 36*(4), 101377.
Ranerup, A., & Henriksen, H. Z. (2020). Digital discretion: Unpacking human and technological agency in automated decision-making in Sweden's social services. *Social Science Computer Review, 1-7.* https://doi.org/10.1177/0894439320980434
SALAR. (2018a). *Automation in case management. Saving time to increase value (Automatiserad ärendehantering. Att frigöra tid för värdeskapande arbete).* SALAR.
SALAR. (2018b). *Invitation for bidding in the public procurement process regarding advanced project leadership and dissemination of experiences in the implementation of a digital co-worker in case management of social assistance [Förfrågningsunderlag. Upphandling av kvalificerade processledartjänster för erfarenhetsspridning och införandeprojekt av digital medarbetare för handläggning av ekonomiskt bistånd].* Förenklat förfarande. SALAR.
SALAR, & PwC. (2018). *Project plan – Qualified process support for dissemination and implementation of a Digital Co-worker in case management in social assistance* (In Swedish): Projektplan – [Kvalificerad processledartjänst för erfarenhetsspridning och införandeprojekt av digital medarbetare för handläggning av ekonomiskt bistånd]. May 2018. SALAR.
Spielcamp, M. (Ed.) (2019). *Automating society. Taking stock of automated decision-making in the EU.* Algorithm Watch in cooperation with Bertelsmann Stiftung, supported by the Open Society Foundations.
Svensson, L. (2019a). Automation – to benefit or ruin? [Automatisering – till nytta eller fördärv?]. *Socialvetenskaplig tidskrift, 3–4,* 341–362.
Svensson, L. (2019b). *The technology is the simple part – Implementing digital automation in case handling in social assistance [Tekniken är den enkla biten. Om att implementera digital automatisering i handläggningen av försörjningsstöd].* University of Lund & Akademikerförbundet SSR.
The Government. (2021). A well-functioning system for elections and decision-making in municipalities and regions [En väl fungerande ordning för val och beslutsfattande i kommuner och regioner]. *SOU, 2021,* 16. The Government.
Trelleborg Municipality. (2015). *Just continue! A way to better quality [Orka fullfölja. Det är en kvalitetsfråga.]* Trelleborg Municipality.

Trelleborg Municipality. (2017a). *Report 2017: First 9 months – Labor Market Board [Delårsrapport 2 2017 Arbetsmarknadsnämnden]*. Trelleborg Municipality.

Trelleborg Municipality. (2017b). *The Trelleborg Model. Automation frees humans to meet other people*! Newsletter, December 5, 2017. [Trelleborgsmodellen. Med automatisering frigörs resurser för det mänskliga mötet]. Trelleborg Municipality.

Wihlborg, E., Larsson, H., & Hedström, K. (2016). "The computer says no" – A case study on automated decision-making in public authorities. In *Proceedings of the 49th Hawaii international conference on system sciences*. IEEE.

Wirtz, B., Weyerer, J. C., & Geyer, C. (2019). Artificial intelligence and the public sector – Applications and challenges. *International Journal of Public Administration, 42*(7), 596–615.

Application of RPA for Cross-Border Business Processes Based on the Example of Intra-Community Supplies

Dominic Alexander Neu, Alessandro Benke, Robert Müller, and Peter Fettke

1 Introduction

The complexity of performing business grows continuously, which is partly caused by an increase in compliance. Tax compliance is a labour-intensive and time-consuming. This applies in particular to the area of value added tax (VAT) compliance. In the EU, transactions between entrepreneurs are often carried out across borders, so-called intra-Community supply. Fraudulent entrepreneurs commit VAT evasion amounting to 60 billion Euros every year by incorrectly claiming input VAT reimbursements (Ismer & Schwarz, 2019). This fraudulent behaviour has led to extensive compliance regulations in this area, which are complex and highly penalised if disregarded (EU-commission, 2017).

Automation technologies such as Robotic Process Automation (RPA) and Artificial Intelligence (AI) are powerful tools for standardisation (Fettke & Risse, 2018). A variety of currently human routine tasks is necessary to fulfil VAT compliance. For this reason, VAT is suitable for implementing automation initiatives. Robotic Process Automation strives in this context since it works on the user interface enabling lightweight integration across the various systems involved (Bal, 2019). In business practice intra-Community supplies demand significant amounts of manual work like collecting and evaluating documents. This compliance work causes friction in intra-Community trade and leaves opportunities for fraudulent activities. Our solution helps to integrate and evaluate information from different sources and thus standardising and automating the process.

D. A. Neu (✉) · A. Benke · R. Müller · P. Fettke
German Research Center for Artificial Intelligence (DFKI), Institute for Information Systems, Saarland Informatics Campus, Saarbruecken, Germany

Institute for Information Systems, Saarland University, Saarbruecken, Germany
e-mail: dominic.neu@dfki.de; alessandro.benke@dfki.de; robert.mueller@dfki.de; peter.fettke@dfki.de

G. Juell-Skielse et al. (eds.), *Service Automation in the Public Sector*, Progress in IS,
https://doi.org/10.1007/978-3-030-92644-1_8

Typical applications for RPA are a combination of manual human-centric work and already automated tasks of existing systems. RPA provides software artifacts, so-called bots or agents, that interact with different systems, enter information, and thus replace repetitive human work (van der Aalst et al., 2018). Automation implies the technical implementation of a process without human or minimal human intervention. The term RPA is sometimes used for artifacts with cognitive components which are additionally capable of replacing human intellectual work (Houy et al., 2019; Viehhauser (2020)). In contrast to the automation of processes in ERP systems, in which processes and the system must be changed, the ERP system is retained by RPA, and existing processes can be automated (Czarnecki & Fettke, 2021, S. 4). Challenges in the tax area exist due to various external stakeholders involved in tax processes, missing system standardisation in cross-border situations causing system and media breaks. This is underlined by the current literature. Classical administration and accounting tasks are tackled through RPA case studies (Cooper et al., 2019; Houy et al., 2019), while taxation processes remain rather unexplored territory. Sala (2020) stresses the importance of tax automation while suggesting starting with repetitive value added taxes and Mezzio et al. (2019) highlight the importance of understanding and analysing tax processes prior to automation.

As a consequence, the following section gives a brief definition of Robotic Process Automation to emphasise the benefits of automation through RPA in the context of Business Process Management (BPM). Then the relevance of accurate process modelling as a transfer format from the business and compliance needs onto the implementation details of RPA robots is discussed. Finally, we present HERAKLIT as an infrastructure to model computer-integrated systems to accomplish this task (Houy et al., 2019, S. 63).

2 RPA: Research Field

Robotic Process Automation has become an increasing trend in research and practice. A broad variety of cases have been presented (Romao et al., 2019; Aguirre & Rodriguez, 2017). RPA is an innovative approach to transform the process execution without changing the underlying application system. First, authors have characterised those systems by the ease of implementation—no or only minimal programming effort—and the simple integration in existing application landscapes through user interfaces. Others emphasise the imitation of tasks that a human would do (van der Aalst et al., 2018). Defining the term RPA is not that easy, as it is more an idea inspired by practice instead of a carefully developed concept. It is a term used for a wide variety of automation techniques.

Increasing the automation of a process through technological progress is not a new idea but a core topic of various business, informatics, and engineering disciplines, that has been discussed for decades. The innovation potential resulting from the interaction of processes and technology was already addressed in the 1990s in the context of Business Process Reengineering (BPR). The core focus of those traditional process automation approaches is an execution of process flows according to mainly static requirements that are either hardcoded into application systems (ERP, CRM, ...) or defined in a Business Process Management System (BPMS). While these systems provide highly reliable integration of many tasks, they imply the costly and time-consuming process of customisation and adaptation to the needs of the individual processes in the enterprise.

The automation of business processes via software robots, on the other hand, starts with another perspective: The manual interaction of humans with existing applications is observed with the goal of imitating this behaviour with a software robot. This bot then takes over the execution-specific activities, which contains the handling of involved application systems through the existing presentation layer. Hence, neither processes nor application systems are changed, and data is only handled via existing applications (Czarnecki & Fettke, 2021).

The functional scope of an RPA system is thereby defined through the selection and description of the process to be automated. Individual activities might range from simple emulation of routine tasks to autonomously reaching complex decisions. Although automation starts with the observation of human actions, proper automation of a manual process does not necessarily mean that all tasks are performed in an equivalent manner (Czarnecki & Fettke, 2021). This makes it difficult to understand and manage the transformation from manual to automated processes. Furthermore, it might be helpful to extend the software robot by further functionalities that could help to improve existing processes or even develop entirely new processes.

The main benefit of RPA lies in the lightweight integration of systems. This opens the opportunity to automate activities, where the economic benefit potential is usually in a range that would not allow a substantial reengineering project. Additionally, the idea of a software robot implies a flexible and agile entity. The composition of different software robots forms a highly dynamic and complex system.

To manage and control the transformation process of manual steps to RPA-bots, we propose the modelling language HERAKLIT. It has been developed to depict the dynamics of business processes and their interplay with information systems while accurately tracking the role of static data elements.

Through the usage of HERAKLIT, the status quo of the business process can be described with the same language as the automated bot-based process. HERAKLIT will be used to model "what" an RPA-Bot will do instead of "how" a bot will execute its actions. The model thus abstracts from the concrete implementation of a bot but is tangible enough to be used as an implementation instruction. As will be shown in Sect. 4.3, the language also allows documenting the execution of bots on an abstract and comprehensible level.

3 Methodological Approach: Modelling RPA-Based Systems with HERAKLIT

HERAKLIT is a general-purpose modelling language based on symbolic Petri nets. HERAKLIT summarises objects and operations on objects of all kinds in a structure: individual objects, documents, abstract objects, services, data, deadlines, invoices with payment terms, composed objects et cetera. The signature is a purely symbolic representation of several structures at the same time. Structures and signatures allow HERAKLIT to model static aspects of business processes. The dynamics and state changes of individual objects are illustrated via the well-founded and proven concept of Petri nets. Lastly, the architecture of larger complexes can be described through the notion of modules. These have interfaces with gates, allowing them to interact with other modules (Fettke & Reisig, 2020). A distributed run is a causal Petri net, containing one event and its predicates for every transition that fired. It therefore describes the occurrence of actions (Reisig, 2013).

These features render HERAKLIT an ideal toolset for modelling RPA-based process automation. The modular architecture allows for modelling complex systems of interdependent bots performing complex tasks. Using distributed runs, it can be shown how information is processed and passed on by and between the bots. HERAKLIT will be used in the context of a design-oriented research approach. It is employed to describe the information architecture and the dynamic of transforming information objects in a precise way.[1]

4 Case Example: VAT, HERAKLIT Model

4.1 VAT Compliance in Intra-Community Supplies

Intra-Community supplies can be exempted from VAT, according to Art. 138 VAT-Directive (2006/112/EC).[2] Depending on the perspective, intra-Community deliveries and intra-Community-acquisitions are differentiated.[3] In simplified terms, the entrepreneur making the supply must fulfill three conditions to qualify for exemption: Confirmation of the actual transport, determination of the entrepreneur status, and the receiving goods are subject to VAT (Amand, 2016, S. 98).

Intra-Community supplies may involve two parties, but they get significantly more complex when several companies are involved. In a chain transaction, for

[1] For additional details, we refer to http://www.heraklit.org/

[2] Council Directive 2006/112/EC of 28 November 2006 on the common system of value added tax, OJ L 347/1.

[3] In this section, we will use the term intra-Community supplies for acquisitions as well as deliveries.

example, the supply is only physically moved between two parties, but the financial transaction is between at least three parties (Dietsch, 2018). For simplicity reasons, this chapter covers only intra-Community supplies between two parties.

The high amount of value added tax poses a substantial risk for entrepreneurs if misjudged. In addition, an intra-Community supply can be revoked by the cause of missing or incomplete documentation. This is especially challenging for small and medium sized companies with non-regular transactions to unknown recipients. For these reasons, the intra-Community supply plays well as a candidate for Robotic Process Automation: high volume, standardised process, well-described decision rules and need for precise documentation.

In the following, the status quo of the verification process as it is usually performed by legal experts is shown. The process can be broken down into four categories of verification steps and the final consolidation of the individual results. The verification steps themselves are executed by a legal expert in an arbitrary sequential order. The documentation of a legal assessment, on the other hand, always follows a strict order to make it more readable for others. To model this process, we therefore differentiate between six different modules. The arbitrary sequential working order of the legal expert is shown in the module "legal experts" (Fig. 1). The module "Merge Verification Results" takes the individual results and merges them sequentially according to the legal best practice (Fig. 2). Whereas the remaining four modules are used to categorise the verification tasks. They do not contain any dynamic behaviour but simply pass the result from the left interface of the module to the right interface. The composition of these modules is shown in Fig. 3.

Although this approach is directly transferable to a bot-based assessment, there are several arguments to split this task into several independent verifications. By reducing the scope of one bot to a smaller process chunk, the maintainability of the systems is increased. This modularisation concept (or micro-services) is well established in software development (van Vliet, 2008). Furthermore, the modularisation of the verification process enables the abstraction of lower-level actions to higher-order activities/subprocesses and thereby enhances the understandability of the system. Finally, by breaking down the work into smaller chunks, the resources can be scaled more precisely to the need of each individual verification task.

Another problem of the process model in Fig. 1 is the detailedness. Although the model might be sufficient as a work instruction for a legal expert, it lacks more detailed instruction on the data involved in the process and the sub-steps necessary to perform each action. The model is not sufficiently concrete to be used as an RPA-Bot construction manual. We, therefore, propose a second, more detailed module (RPA Solution) that clearly outlines the required data for each step and provides actions of lower granularity that are ready to be implemented algorithmically. The module can substitute the "legal experts" module in Fig. 3 while the verification tasks and the merging of the results remain identical.

Since the law is a science in which much depends on contextual circumstances, we must presuppose some premises so that the algorithmic implementation corresponds to the legal examination procedures. Some of these premises are general

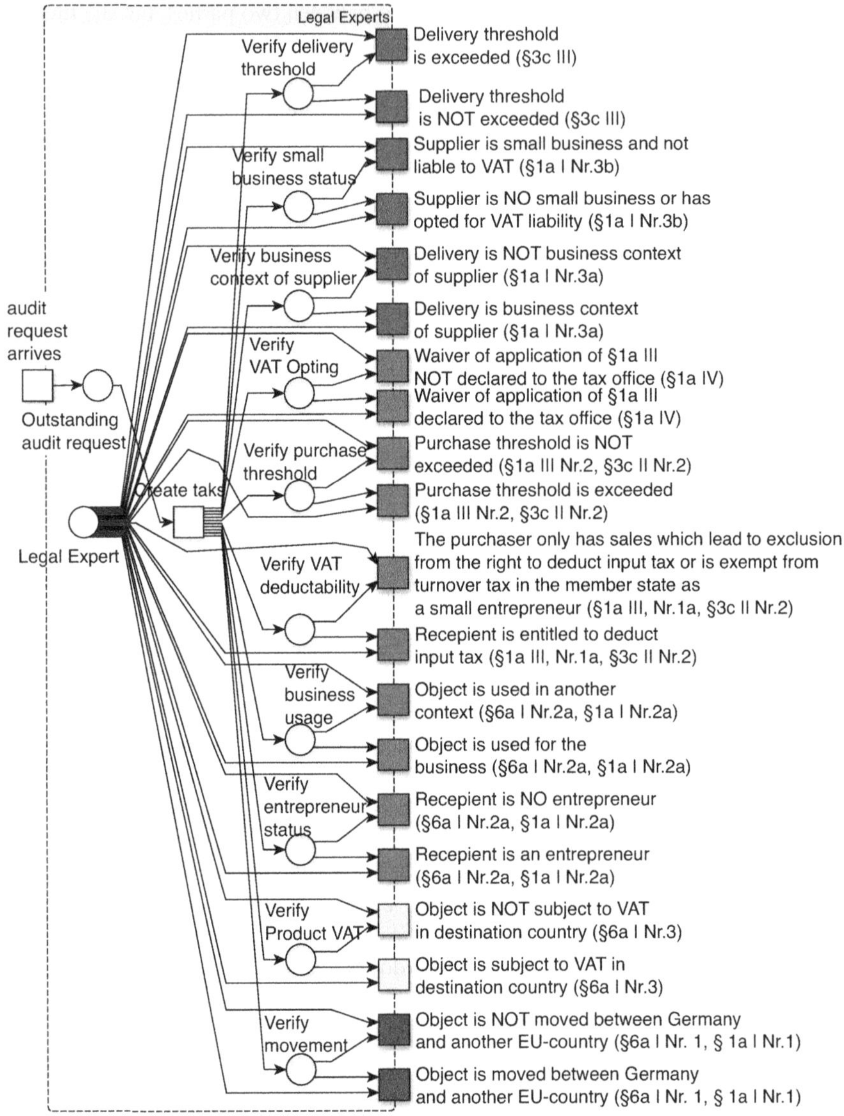

Fig. 1 Module Legal experts with inner dynamics of verification steps

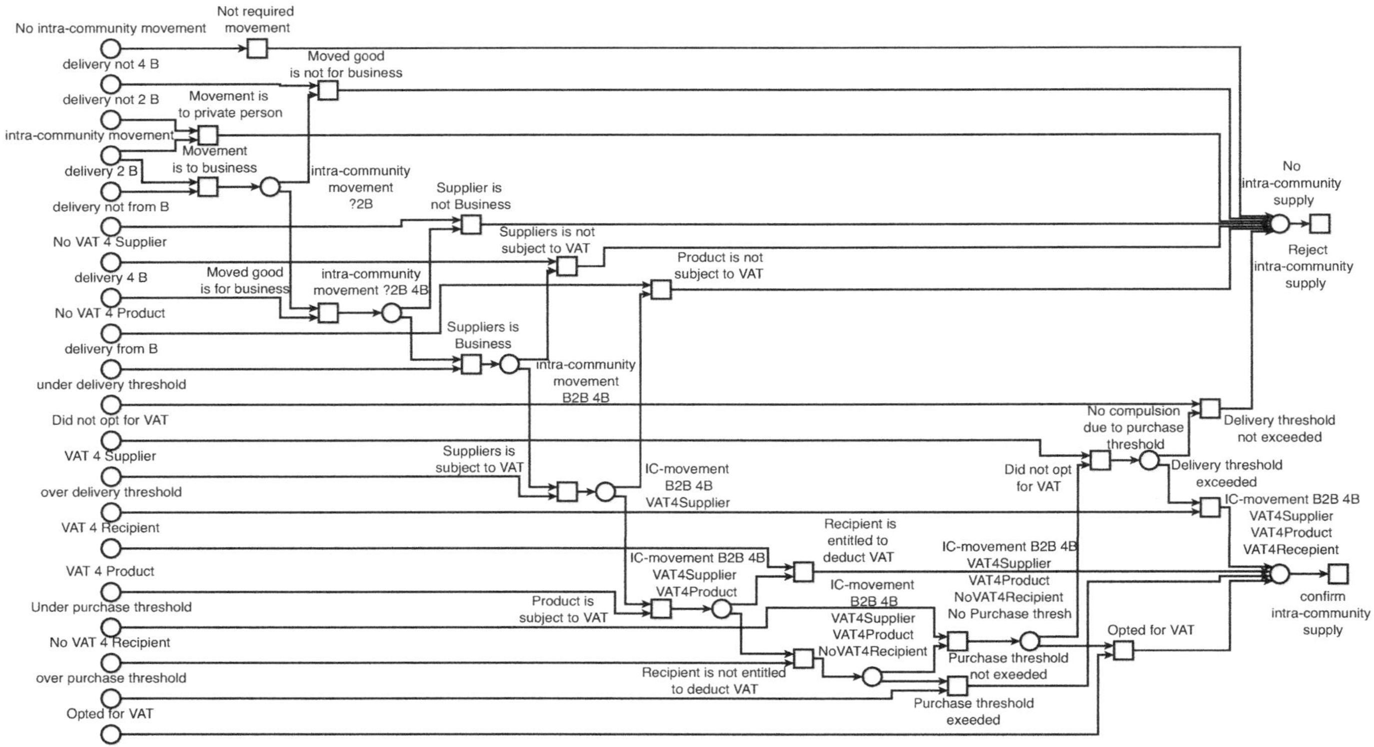

Fig. 2 Sequential Merging of results for the documentation (2 and 4 are used as abbreviations for "to" and "for", respectively. Furthermore, B is the abbreviation for "business")

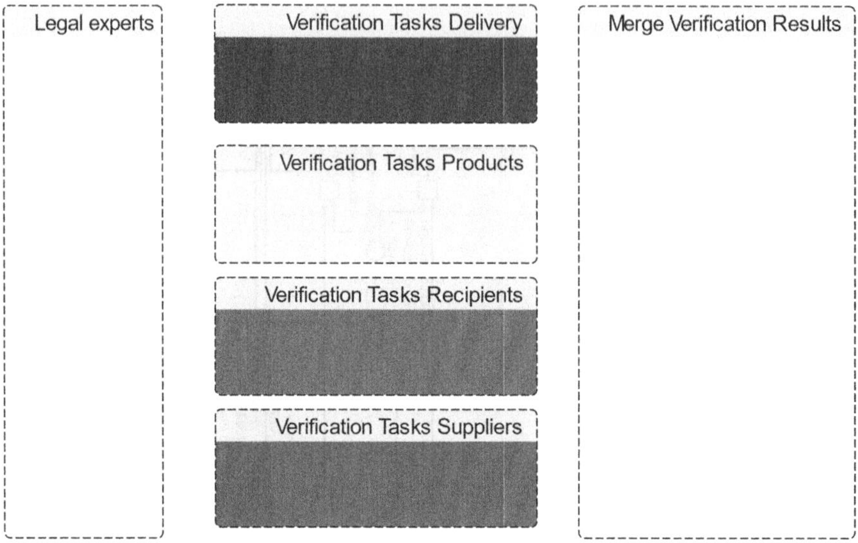

Fig. 3 Module composition with legal experts

practices accepted by the courts; others are enterprise-specific and need to be evaluated individually by the applicant.

4.2 *RPA Solution Model*

To model the legal assessment of intra-Community supplies, the modelling language HERAKLIT is used. It amalgamates the detailed description of static data involved and the dynamic behaviour of a process. In addition, its strict formalism simplifies the transformation into an algorithmic implementation.

To start, the static data objects involved in the process are enumerated and composed. Together they form the signature of the model (see Table 1):

Through this signature, the required static elements are further characterised without a reference to a specific data format. It should be noted that the sort "Order" can either be an invoice (for the recipient of an intra-Community supply) or a purchase order (for the supplier). Although the logical structure of an invoice is prescribed in a more detailed form (e.g. invoice number, tax rate etc.), the signature is reduced to the data elements necessary to perform an intra-Community-supply verification. The invoice is assumed to conform with additional formal requirements from other regulations.

Following the static elements, the composition of each submodule is presented in Fig. 4. An order can arrive at any time, and when it does, it is split into several smaller data objects, each containing only the necessary information to perform the following verification. This is done in the module Split Tasks (Fig. 5). Afterwards,

Table 1 Signature of the model

Base sorts		**Variable names**
Legal entity		enty,enty_1, enty_2
Product		prod, prod_1
Product category		categ
VATIN		vatin
Country		cntry, cntry_1, cntry_2
Status		stat, stat_recv, stat_supp, stat_deliv, stat_prod
Derived sorts		
Company	Legal entity × country	comp
Partner status	Company × status	–
Receiver	Company × VATIN	recv
Order positions	P(product)	opos
Order	Company × receiver × order positions	ordr
Functions		
Cat: Product -> product category		

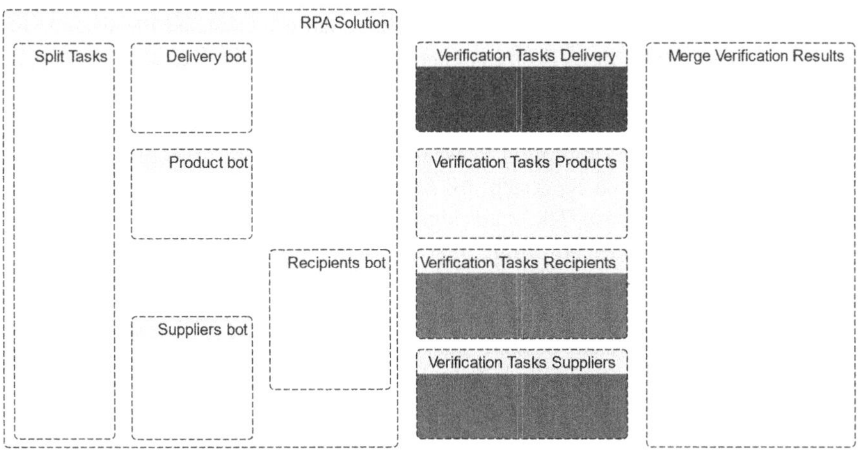

Fig. 4 Composition of Module "RPA Solution"

each individual verification is performed in a separate module. The modules Verify receiver and Verify supplier investigate the required status of the companies involved and share a common data object, the Partner Status, but other than that, they are independent. The module Verify delivery checks for sufficient documentation of the shipment to the foreign EU country. The module Verify products ensures that the individual products are subject to the VAT in the destination country. At last, the module Merge Verification results merges all four outcomes and proposes one final assessment.

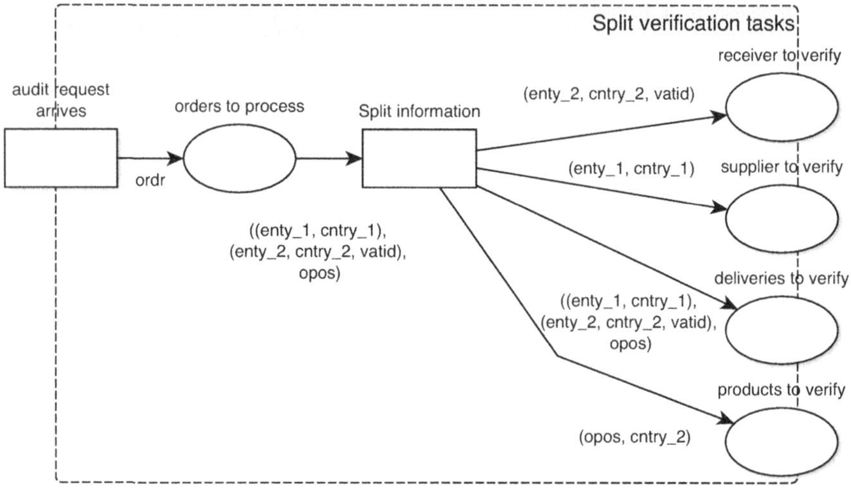

Fig. 5 Module "Split Verification Tasks"

The assessment of the receiver is split into three consecutive steps. First, the specification of the VATIN is verified. By this, the receiver signals that this purchase is on behalf of a company liable for turnover tax, and it is intended for usage in his business.[4] Afterwards, the VATIN is checked for its validity, which, in Germany, can be done through the portal of the Federal Central Tax Office.[5] Although an invalid VATIN is handled equivalently to a missing VATIN, further inquiries to the receiver might be an appropriate escalation step before rejecting the intra-Community supply. As a final step, the status of the receiver is retrieved from an internal customer database. The status signals whether a company is excluded from intra-Community supplies according to some legal norm. Since this status is somewhat unlikely to change throughout the year, it is updated periodically in the database and not checked individually for every invoice (Fig. 6).

This premise will also depend on the company in focus. For example a manufacturing enterprise will likely have just a handful of clients in fixed countries. In this case, the partner status might be updated yearly in a spreadsheet or the ERP system. An online retailer, on the other hand, would encounter new clients from different countries daily. In this setting, the verification of the partner status itself should as well be modelled and automated.

The verification of the supplier follows a similar pattern to that of the receiver, with the distinction that a supplier does not specify a VATIN. Therefore, it is only the status in the database that is verified.

The verification of the delivery is divided into two steps. Firstly, the requirement of delivery between two distinct EU countries is verified. This is achieved by

[4] Art. 43 Para. 2 MWSt SysRl.

[5] https://evatr.bff-online.de/eVatR/index_html

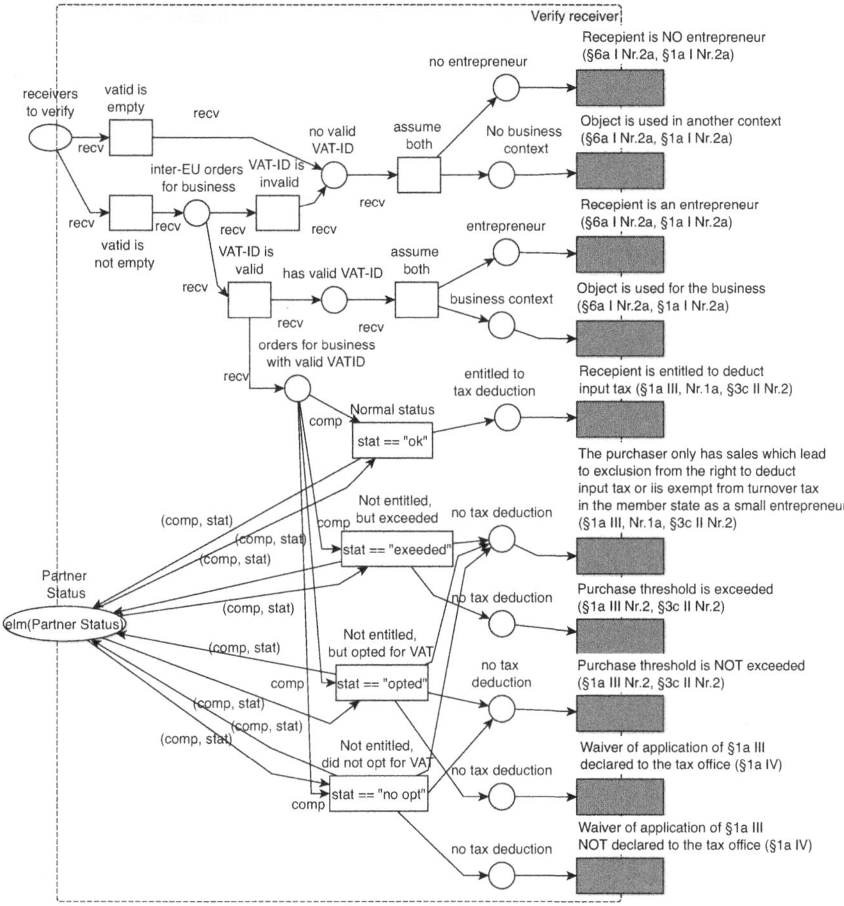

Fig. 6 Module "Verify Receiver"

comparing the two countries of each partner involved. For the delivery verification, it is not sufficient that two EU companies are involved. The proper execution of the product movement from one country to another must be documented via confirmation of arrival. Therefore, as a next step, the existence of confirmation of arrival relating to this order is ensured (Fig. 7).

National law and court decisions strongly influence the exact formal requirements of a valid confirmation of arrival (Langhein, 2017, S. 144 ff.). The detailed validation of a CoA is therefore abstracted in this context. However, in a practical setting, the verification of the CoA itself should be modelled as an inner module containing the detailed verification steps.

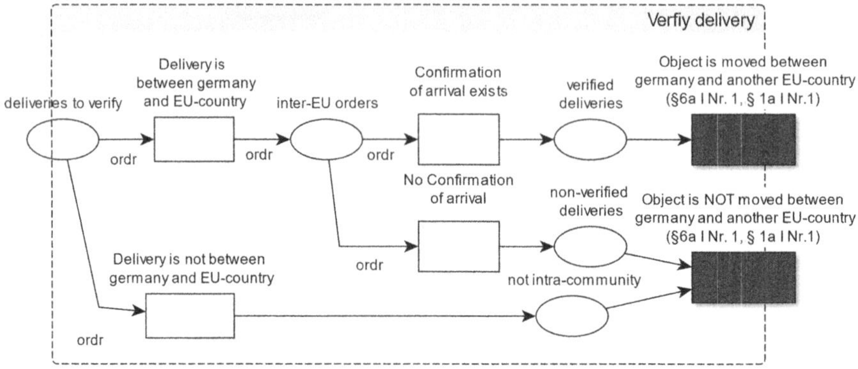

Fig. 7 Module "Verify delivery"

The remaining verification task checks for each product if it is subject to the VAT in the destination country. Although this step is part of our model, the explicit implementation will depend on the required rigorousness defined by the company and the products in question. Many exceptions differ throughout the different member states of the EU. Depending on the products sold by the company, several approaches might be recommendable: assume the product is always subject to VAT, infer the result from the product category or hand over each case to a human worker and process the result. In our model, we infer the tax liability from the product category (Fig. 8).

Like the invalid VATIN, another escalation step to handle mixed orders might be recommendable, but mixed orders are treated as non-VAT liable for simplicity reasons.

After each verification has been executed concurrently, the intermediate results are merged into a final one in the module Merge Verification results. This is basically a logical join on all the possible combinations of individual module outcomes and is identical to the merging performed by a legal expert.

4.3 Intra-Community Supply: Verification Example

To present the necessary steps in the legal assessment of VAT exemption, the verification schema is presented by one exemplary case. The evaluation is generally based on the German VAT act.[6] However, not only the national legal situation is

[6]Value Added Tax Act (Umsatzsteuergesetz) in the version of the notice of 21. February 2005 (BGBl. I S. 386), which was last changed through Article 3 of the Act of 10. March 2021 (BGBl. I S. 330).

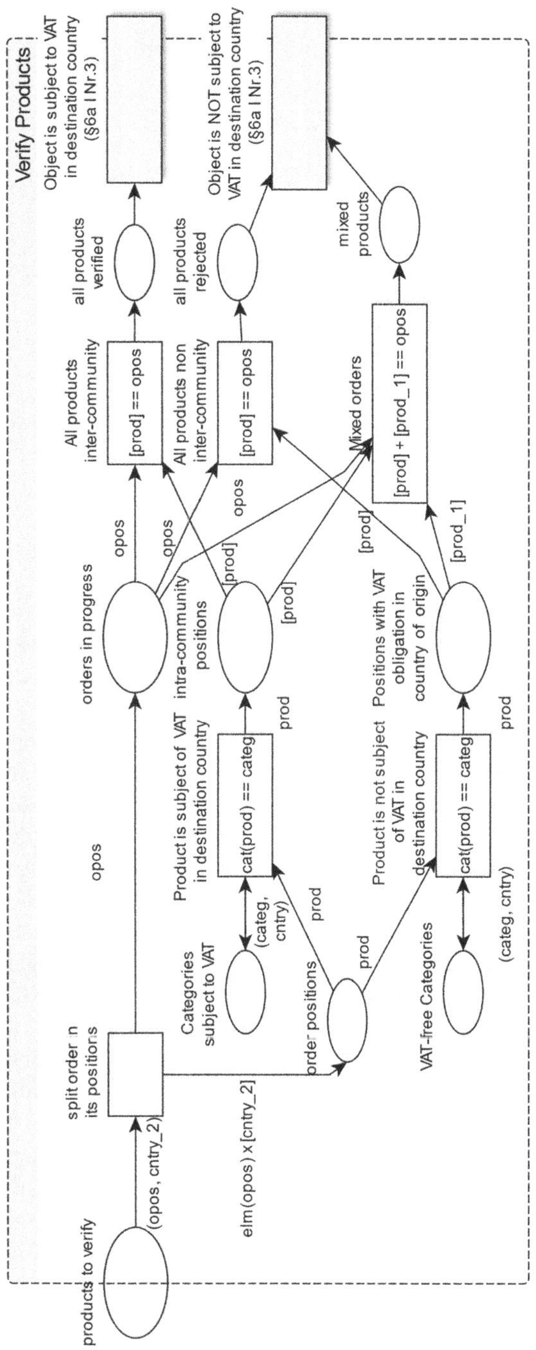

Fig. 8 Module “Verify Products”

Table 2 Initialised Signature

Base sorts	Variable names
Legal entity	[MedicalSupplies SE, MyCorp GmbH]
Product	["cough syrup"]
Product category	[medicine]
VATIN	[DE314159265]
Country	[Germany, Spain]
Status	[excluded, "not excluded"]
Derived sorts	
Company:	[(MedicalSupplies SE, Spain), (MyCorp GmbH, Germany)]
Partner status:	[((MedicalSupplies SE, Spain), not excluded), ((MyCorp GmbH, Germany), not excluded)]
Receiver:	((MyCorp GmbH, Germany), DE314159265)
Order positions:	[[["cough syrup"]]
Order:	[((MedicalSupplies SE, Spain), ((MyCorp GmbH, Germany), DE314159265), ["cough syrup"])]
Functions	
Cat: Product -> product category	Cat(cough syrup) = medicine

decisive, but also the legal situation in the other Member States (e.g. the delivery and the acquisition thresholds are specific to each country). Furthermore, the national laws and court decisions further influence the formal requirements concerning the invoice and the assessment documentation.

In this case, the Spanish Medical supplies SE delivers cough syrup to the German MyCorp GmbH. Cough sirup belongs to the product category of medicine and is subject to the VAT in Germany. Moreover, the Germany Company supplied its valid VATIN when placing the order. This action signals the Medical Supplies SE that MyCorp GmbH is liable to the VAT in Germany and that the cough syrup will be used in the company's business context. MyCorp GmbH and medical supplies SE are known business partners; therefore, it has been verified for each company that they have not been excluded from vat exemption through some legal norm.

The statics of this verification example are shown in Table 2, where each sort is initialised with its corresponding data elements. The events of the verification process are aligned with their connected data objects in the distributed run of the schema. Figure 9 shows the distributed run for this example and also constitutes a precise documentation of the verification process and its result.

Contrary to classical logs, there is no total order in the events. Usually, a timestamp is used to create a total order of the events. In this case, the order is only defined through the casual relationship of events. In Fig. 9 it can be seen, that the individual verification steps (turquoise, dark cyan, olive, purple) are independent and concurrent, whereas all verifications need to finish before the final result can be produced (brown events). To get a more detailed look at the event inscriptions, Fig. 10 depicts an excerpt of the distributed run, containing the verification steps that relate to the product itself.

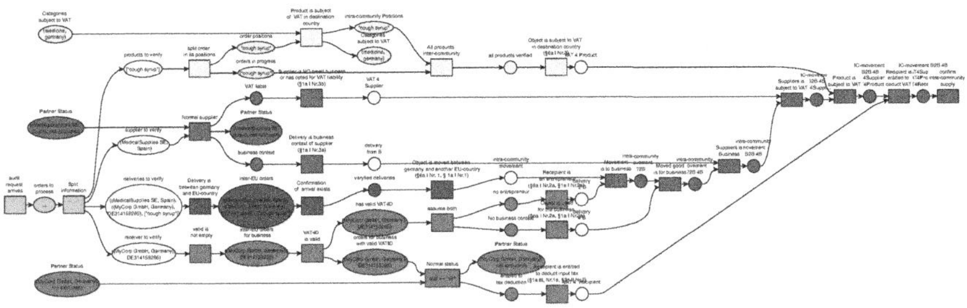

Fig. 9 Distributed run as event log (Inscriptions are not readable and only added for completeness. The general structure is of interest)

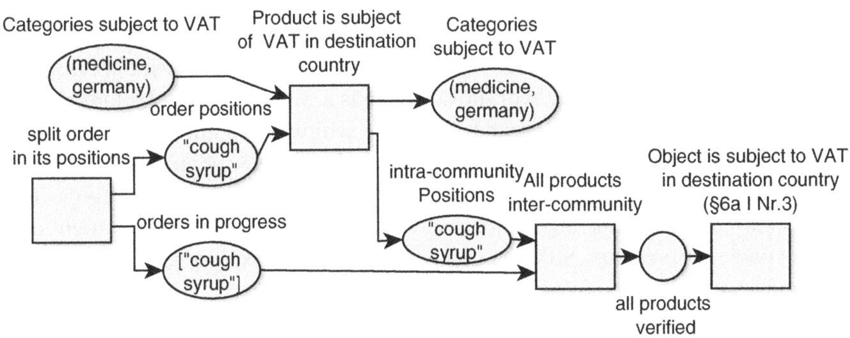

Fig. 10 Excerpt of distributed run

For example the event "Object is subject to VAT in the destination country (§6a I Nr.3)" shows that the bot verified the specified norm. By traversing the distributed run to the left, the reason for this decision becomes evident: The product in question (cough sirup) belongs to the product category medicine and is therefore subject to VAT.

In conclusion, HERAKLIT is capable of modelling the RPA system and documenting the execution of processes. The acyclic graph allows extracting causal relationships between events and data objects, thereby enhancing the explainability and understandability of RPA-Systems for business users and legal experts.

4.4 *Process Automation Through Multi-Agent RPA*

The model in Sect. 4.1 describes the distributed process of intra-Community supplies. Although the dynamics and the involved data objects are defined, the model is technology agnostic. The following section will describe the transfer of this model onto specific multi-agent RPA implementations.

Research has shown that the division of larger process automation into smaller fragments reduces maintenance costs and improves the whole system's understandability. This is also reflected in the HERAKLIT model. For an initial understanding of the assessment process, only the modules and their interfaces are utilised. The inner logic of each process is abstracted. This representation can be directly transferred to the RPA implementation. Each module can be implemented as a separate RPA-Agent, and the interfaces depict the communication channels and the expected information exchange.

A principal design question when constructing modular RPA automation is whether to connect bots via choreography or orchestration. In an orchestrated execution, a central party like a process engine or a workflow management system communicates with each bot individually and controls their collaboration. (Mendling & Michael, 2008) Workflow engines are centralised services that were designed to orchestrate processes within an organisation. The idea to control process step executions in multiple systems through one centralised system suffers from the following problems: A central workflow engine is a Single Point of Failure (SPOF) and hard to scale to large, distributed enterprise settings. In addition, they require a single agreed-upon process model. Although this might be possible in intra-enterprise processes, this becomes highly unlikely in inter-enterprise settings. This is especially true for value added tax assessments since a single invoice might only involve two entrepreneurs. Still, the company assesses massive amounts of invoices with numerous suppliers and clients.

The choreographic approach, on the other hand, imposes fewer restrictions on the stakeholders. The stakeholders instead agree upon a model describing how information is exchanged. Each stakeholder can independently decide on the internal architecture of the services he contributes as long as they implement the underlying choreography model (Mendling & Michael, 2008). This approach is more modular and robust to changes.

The HERAKLIT Modell imposes a choreographed solution where the gates at each module interface represent the information channels. Each module can be implemented as a separate RPA-Bot, and the interface defines the required communication with other parts of the system. Furthermore, this architecture can be scaled to other tax-related automation without the need for one unified model. Therefore, Aalst describes RPA as "BPMS for the poor" since it is a loosely coupled and cheap connection of IT systems with a business process perspective (van der Aalst, 2021).

5 Discussion

This case study showed the ability of HERAKLIT to describe the logic of an intra-Community verification process. We were able to model the status quo of the legal experts and the automated process with RPA-Bots. Through modelling the verification process with legal experts, the complexity of taxation automation became visible. Even though multiple simplifying assumptions were made, the model complexity had to be reduced through modularisation.

The contribution focuses on the modelling perspective of complex interdependent systems of bots to automate entire workflows and legal decision-making. The implementation is not in the scope of this contribution.

Splitting the verification process into smaller, independent bots not only enhances the understandability but also leverages the benefits of concurrent computing and modularisation:

- The bots are easier to maintain since the complexity of one bot is reduced.
- For semi-automated processes, modularisation allows for more elaborate tasks to be left to human workers.
- Hardware provisioning can be adapted more adequately to the requirements of each bot. For example neural network-based optical character recognition in a document might need GPU acceleration, whereas classical front-end automation demands fewer resources.

In this context, it should be mentioned that a sequential control flow that breaks when encountering KO-criteria might be more resource-efficient in obtaining a result since the execution would stop at an earlier point in the verification process. On the other hand, a concurrent solution ensures that each verification task is completed and that the resulting documentation is always comprehensive.

Moreover, the graph-based documentation of the process events enables business users, tax office officials and legal experts to extract causal relationships. Although, the distributed form of legal assessment might seem unusual to legal experts. This is primarily due to the fact that a verification assignment would not be split into such small tasks. The communication overhead to pass a minimal verification task to a co-worker would exceed the time to perform the actual verification. For example the time requirements for writing an email with the assignment: Please check the following VATIN is equivalent to entering the VATIN into the form of the Federal Central Tax Office. This does not apply to RPA-Bots communicating with each other.

Furthermore, we believe the presentation of RPA-execution logs as distributed runs which are based on the process model developed by business users, dramatically enhances the understandability of such systems for non-technical employees.

By using individual transitions for each verification step, we were able to link individual bot activities to their corresponding legal norm. This is another enhancement of the documentation since each activity is linked to its semantics. Through this link, tax office officials are enabled to validate the semantic equivalency between the bot actions and the intended meanings of the legal norm.

Unfortunately, to create an RPA implementation that fully automates the verification process, further aspects of the model need to be specified. As mentioned earlier, the validation of the confirmation of arrival should be modelled in itself, and the maintenance of the partner status should be considered for automation. Through the modularisation capabilities of HERAKLIT, these models can be integrated seamlessly into the existing work. The modularisation furthermore allows to switch individual components: the module "product verification", for example can have different internal dynamics depending on the needs of the company. If multiple reference model alternatives are presented by academia, the company only needs to insert the module with congruent premises.

6 Outlook

At first sight, verifying intra-Community supplies constitutes a perfect candidate for automation: high volume throughput, standardised decision rules, extensive documentation requirements and millions of European enterprises as potential beneficiaries.

The difficulties arise in the assessment of each individual case. Although the general decision rules are specified only by a handful of norms, they are refined and detailed in a wide range of tax directives, implementation ordinances and regional court decisions. This requires the automation model to differentiate between a multitude of cases. In addition, legal norms require a factual assessment which means a legal expert decides if a rather open formulation is applicable in a complex situation. This requires contextual information and experience, which are hard to leverage in current algorithmic implementations. Automations are at their best when an unambiguous measurand is constrained using a fixed domain of values. Therefore, in our opinion, the task of auditors will predominantly be to define heuristics that map an unambiguous measurand to an acceptable range of values. These heuristics must be defined in such a way that their results are congruent with the auditors' common assessments in many circumstances.

The difficulties in automating the entire workflow shown throughout this case study support the effectiveness of Heraklit as an intermediary modelling language between subject-matter experts and explicit implementations. By conceptualising the architecture, the statics, and the dynamics of an RPA solution, we uncovered unsolved questions that need to be addressed prior to the execution phase of an RPA project. Additional research could build upon this insight and examine further automation cases in public administrations.

In the next step, we plan to extend our model to cover a wider range of cases and implement a prototype based on this reference model. The open research questions are whether the Heraklit model is expressive enough to create an RPA implementation based on it and if the distributed run serves as an enhanced and easily comprehensible execution log.

References

Aguirre, S., & Rodriguez, A. (2017). Automation of a business process using robotic process automation (RPA): A case study. In *Workshop on engineering applications* (pp. 65–71). Springer.

Amand, C. (2016). The impossible proof of intra-community supplies of goods. *International VAT Monitor, 27*, 98.

Bal, A. (2019, April 15). *Indirect tax compliance: Process automation or process transformation?* (pp. 255–258). Tax Notes International.

Cooper, L. A., Holderness, D., Sorensen, T. L., & Wood, D. A. (2019). Robotic process automation in public accounting. *Accounting Horizons, 33*(4), 15–35.

Czarnecki, C., & Fettke, P. (2021). Robotic process automation. Positioning, structuring, and framing the work. In I. C. Czarnecki & P. Fettke (Eds.), *Robotic process automation. Management, technology, applications* (pp. 3–24). De Gruyter.

Dietsch, D. R. (2018). *Umsatzsteuer 4.0—wie Blockchain grenzüberschreitende Reihengeschäfte transparenter machen könnte* (p. 813). MwStR.

EU-commission. (2017)., Towards a single EU VAT area—Time to act. Amended proposal for a council regulation amending regulation (EU) no 904/2010 as regards measures to strengthen administrative cooperation in the field of value added tax, COM 2017 706 final. Brussels.

Fettke, P., & Reisig, W. (2020). Heraklit—die erkenntnistheoretisch motivierte Modellierung rechnerintegrierter Systeme.

Fettke, P., & Risse, R. (2018, July 27). *Blockchain: Wird eine sog. "Real time" tax compliance möglich?* (pp. 1748–1755). Der Betrieb.

Houy, C., Hamberg, M., & Fettke, P. (2019). *Robotic process automation in public Administrations. Digitalisierung von Staat und Verwaltung. Fachtagung Verwaltungsinformatik (FTVI) und Fachtagung Rechtsinformatik (FTRI)* (pp. 62–74). Gesellschaft für Informatik (GI).

Ismer, R., & Schwarz, M. (2019). Combating VAT fraud through digital technologies: A reform proposal. *International VAT Monitor, 30*(6), 240.

Langhein, N. (2017). *Umsatzsteuer und Steuerplanung*. Springer Fachmedien.

Mendling, J., & Michael, H. (2008). From WS-CDL choreography to BPEL process orchestration. *Journal of Enterprise Information Management, 21*, 525–542.

Mezzio, S., Stein, R., & Stein, S. (2019). Robotic process automation for tax. *Journal of Accountancy, 228*(6), 61–63.

Reisig, W. (2013). *Petri nets. Modeling techniques, analysis methods, case studies*. Springer.

Romao, M., Costa, J., & Costa, C. J. (2019). Robotic process automation: A case study in the banking industry. In *2019 14th Iberian conference on information systems and technologies (CISTI)* (pp. 1–6). IEEE.

Sala, M. (2020). *A road map for tax function automation* (pp. 1–6). Tax Adviser.

van der Aalst, W. M. (2021). Process mining and RPA. In C. Czarnecki & P. Fettke (Eds.), *Robotic process automation—Management, technology, applications* (pp. 223–239). De Gruyter.

van der Aalst, W., Bichler, M., & Heinzl, A. (2018). Robotic process automation. *Business and Information Systems Engineering, 60*, 269–272.

van Vliet, H. (2008). *Software engineering: Principles and practice*. Wiley.

Viehhauser, J. (2020). Is robotic process automation becoming intelligent? Early evidence of influences of artificial intelligence on robotic process automation. *Business Process Management: Blockchain and Robotic Process Automation Forum, 393*, 101–115. https://doi.org/10.1007/978-3-030-58779-6_7

Part IV
Implementation Challenges of Public Sector Service Automation

Enhancing Routine Capability Through Robotic Process Automation in the Public Sector: A Case Survey

Evrim Oya Güner, Shengnan Han, and Gustaf Juell-Skielse

1 Introduction

Organizations in the public sector are called upon digitalization of their services, processes, and practices to meet better the needs and requests from people, private business organizations, and many other stakeholders in the society (Schou & Morten, 2018). In recent years, robotic process automation (RPA) has been increasingly adopted in the public sector to digitalize public services, work routines, and processes. RPA is the automation of service tasks through software configuration to perform repetitive and low value-adding work previously done by humans (Willcocks et al., 2015). The configured software is often called "robot" mimicking humans' actions to perform the work and interact with the information systems (Jimenez-Ramirez et al., 2019). Supported by AI capabilities, such as machine learning of unstructured data and images and natural language processing, RPA is becoming more intelligent to complete complicated tasks (van de Weerd et al., 2021). RPA, by extension of business process management practices, is being associated with the digitalization of public services (Houy et al., 2019; Lindgren, 2020).

Previous research investigated the adoption and use of RPA in the public sector (i.e., Dias et al., 2019; Goday-Verdaguer et al., 2020; Houy et al., 2019; Ranerup & Henriksen, 2019, 2020). These studies conclude that the use of RPA improves public service processes as it shortens time to respond, reduces human errors, and increases quality and compliance. However, we have not understood how these improvements

E. O. Güner (✉) · S. Han
Stockholm University, Stockholm, Sweden
e-mail: evrim.guner@dsv.su.se; shengnan@dsv.su.se

G. Juell-Skielse
University of Borås, Borås, Sweden
e-mail: gustaf.juell-skielse@hb.se

G. Juell-Skielse et al. (eds.), *Service Automation in the Public Sector*, Progress in IS,
https://doi.org/10.1007/978-3-030-92644-1_9

in service processes and practices are generated through the practices constituted from RPA-enabled new routines and capabilities. Therefore, the primary purpose of this study is to contribute to the body of knowledge on RPA by identifying changes in routines that stipulate necessary capabilities to advance public service processes and practices. Thus, the question "How does RPA as routine capability advance service processes and practices in the public sector?" guides the investigation in this study.

To obtain an integrative view of how RPA advances process and service practices in the public sector, we adopt "technology as routine capability" (Swanson, 2019) as the theoretical lens to analyze and interpret the cases that are currently published in academic literature.

The routines, from one perspective, include "know-how" and are essential to capabilities with their continuous adaptation (Feldman & Pentland, 2003; Swanson, 2019). As routines are constitutive elements of broader practices, decomposing RPA enabled practices and zooming in within the routines help us to unfold the RPA related changes/improvements in routines, practices, and processes. Technology as routine capability puts the routines forward and provides a useful analytic lens for studying the interrelation among technology, routine, capability, and practice. The distinctive theoretical element of this perspective is that routines are integral to technology itself. The capability is associated with these technology-enabled new routines in which human practices are advanced. As Swanson puts it (2019, p. 1008), "the key here is the view of practice as constituted from routines, facilitating the examination of whole practices in terms of routines." Further, routines are considered as patterns of action not only related to people but also machines (Pentland & Feldman, 2008; Swanson, 2019).

Considering that RPA can potentially replace the human worker or change an entire process, both human and machine routines are in the scope of routine capability perspective. Technology as routine capability provides a meta-analytic framework and allows us to accumulate insights from multiple case studies that "both illuminates and motivates" research on RPA change (Swanson, 2019). Our data sample is extracted from the recent case studies of RPA adoption and implementation in public organizations (see Appendix). The study results demonstrate that RPA as routine capability advance service processes and practices in public organizations.

This chapter contributes to digital government research with a deeper understanding of how RPA changes and cultivates routine capability and advances service processes, and practices in the public sector. In addition, we apply and critically examine technology as routine capability as the analytical framework for understanding how RPA advances public service practice (Swanson, 2019). To that end, in Sect. 2, we discuss ex-ante literature of RPA adoption in the public sector and introduce the theory of technology as routine capability. In Sect. 3, we present the method that we used for the case survey and analysis. In Sect. 4, the cases are analyzed through the lens of technology as routine capability to provide empirical evidence of how RPA advances public service practice and what changes RPA as routine capability in the public organizations. Finally, in Sect. 5, we discuss insights in relation to RPA research, our critical thoughts of the perspective of technology as routine capability, and present our conclusions.

2 Research Background

The research background introduces research on robotic process automation, the recent literature of RPA in public sector and technology as routine capability framework.

2.1 *Robotic Process Automation*

RPA is one of the technologies used in the automation of work processes to improve process efficiency. It can be described as the automation of structured processes executing repetitive, rule-based, high-volume, and routine tasks by configuring the software to perform the work previously done by humans (Willcocks et al., 2015). The configured software is often called "robot," mimicking humans' actions to perform the work and interact with the information systems (Jimenez-Ramirez et al., 2019). Each "robot" equals a single software instance (Penttinen et al., 2018) performing the task on the user interface by executing the process interactions based on the predefined rules through the presentation layers (Cewe et al., 2018).

The main benefits associated with a successful RPA implementation are increased process speed and productivity, error reduction (Aguirre & Rodriguez, 2017; Ratia et al., 2018; Suri et al., 2017), and quality improvements (Jimenez-Ramirez et al., 2019; Ratia et al., 2018). Further, by freeing up the employees from high-volume, non-value-adding tasks, RPA supports employees' reallocation to more critical and value-adding tasks.

RPA, in its emergence, was promoted as a suitable technology to automate repetitive and routine tasks. However, later versions of RPA integrated with AI technologies allow for more complex processes to be automated with RPA. While the early applications of RPA aimed at automating the processes with structured data, the combination of RPA and AI expanded the application fields of RPA to the execution of less-defined tasks, which require the processing of semi-structured and unstructured data (van de Weerd et al., 2021).

2.2 *RPA in the Public Sector*

The literature shows that RPA adoption is increasing among public organizations, as indicated by the growing number of published research in the last 2 years (i.e., Dias et al., 2019; Goday-Verdaguer et al., 2020; Houy et al., 2019; Ranerup & Henriksen,

2019, 2020). Ranerup (2020) reports that around 16 municipalities in Sweden implemented RPA in social work practices in 2019. It is projected that the number of municipalities that adopt RPA will increase due to its promotion by RPA vendors and policymakers (ibid).

Compared to traditional IT projects, RPA offers ease of deployment and use in a shorter time frame with a faster return on investment (ROI) and thus supports digitalization endeavors in public organizations (Söderström et al., 2021). More particularly,

RPA supports the fulfillment of digital government by improving the efficiency, quality, and transparency of the public services and thus can initiate further automation of routine tasks and processes of public organizations (Lindgren, 2020). While the automated processes can provide more efficient and transparent public services, balancing the public values and benefits of new technologies can be challenging (Smith et al., 2010). Ranerup and Henriksen (2019) addressed this challenge and investigated RPA enabled decision-making from the perspective of the public sector values. Their results indicate that RPA enabled decision-making reduced the cost of social assistance services and also contributed to the public sector value positions- professionalism, efficiency, and service.

The use of RPA in public organizations has also implications associated with the civil servants. For example, the implications of performing knowledge-work by humans and machines together in organizations where AI is in use (Dias et al., 2019) and the tradeoffs of automation (Söderström et al., 2021) such as employee lay-offs (Eikebrokk & Olsen, 2020).

2.3 Technology as Routine Capability

By revealing the interplay between technology, routine, capability, and practices, Swanson (2019) theorizes routine as integral to technology itself and capability associated with device (technology)-enabled (new) routines by which human practices are advanced. He terms this perspective as "technology as routine capability." Technology as routine capability relates technology to peoples' use of devices through their routines, rather than with the devices themselves. Swanson (2019) contrasts to technology as device perspective, which views technology as a "means to fulfil a human purpose" and argues that "device-enabled routines constitute technology" through the capabilities generated in human practices.

Technology as routine capability framework brings routines to the foreground of the analysis. Routines are described as patterns of action by people as well as by machines. Swanson (2019) argues that the affordances of devices (technologies) are mostly received through the routines (affordance built-in routine performance) which employ the devices (device as routine component). Routines are constitutive components of broader practices and executed in coordination and particular social order (routine as practice component). Practices are constructed from the purposeful collection routines that provide needed capabilities to advance the practices (ibid). In this reasoning, Swanson (2019) associates capabilities with practices and argues that "technology as routine capability serves to advance human practice."

Advancing human practice emerges in multiple dimensions with regards to improvement, such as "its economics, social acceptance, or politics" (Swanson, 2019, p. 1011). Such improvements include but are not limited to "simplifying it or otherwise reducing its costs, for instance, or finding new outlets for it, expanding its presence, or increasing its appeal, making it more enjoyable, or obtaining favoured social treatment for it, or improving its reputation" (Swanson, 2019, p. 1011).

Swanson (2019) suggests three levels in advancing practices: individual, organizational, and social. The individual level focuses on human practice, which is improved by the new routine capabilities provided by technology. The organizational level is about gaining new or improving existing routine capabilities of the organization through technology. The social level is concerned with the whole society covering industries, professional services, professions, and other units of society advanced by the new routine capabilities.

As can be seen, Swanson (2019) associates capabilities with practices. Suggesting that "technology as routine capability serves to advance human practice," Swanson (2019) concludes that advancing human practice is the driver of change in technology. In advancing practices, four basic modes of change are identified: (1) design, in creating new devices and routines, (2) execution, in operating devices and performing routines, (3) diffusion, in spreading devices and routines to a population's members, and (4) shift, in adapting devices and routines to change among a world's practices. From the perspective of routine capability, change in technology by design applies design cycle procedures to routines instead of devices (Swanson, 2019). However, the design of a routine can only be complete when it is entirely performed (ibid.). In change by execution, practices can be advanced through performance and reformed with each performance. In change by diffusion, advancing the practice requires new adopters to achieve new routine capabilities. In change by shift, emphasis is given to prosperity and persistence of continuous shifts in advancing the practices through adoption and recreation of the routines and practices.

With regards to the application of the framework, Swanson (2019) suggests that a particular practice can be analyzed by "zooming in" and by "zooming out," insights from multiple related practices can be aggregated. "Zooming in" can be conducted having a single technology in a single organization as the unit of analysis. By focusing on routine capability, we aim to study individual and organizational practices, how they are constituted from routines, and how changes are made through routine capabilities that build multiple practices within the organization. "Zooming out" is used to investigate how technology, as routine capability, is forged in advancing practices in whole industries and professions inclusively. Swanson (2019) proposes three types of studies: (1) studies of technologies whose development paths are intertwined across multiple practices; (2) studies with a substantial historical dimension; and (3) studies that relate technology change to broader ecological shifts and transformation of practices in terms of how they compete and cohere.

3 Method

Case survey is a suitable method to overcome the problem of generalization from a single case study by aggregating from several case studies and identifying the patterns across different case studies in an inexpensive and resourceful way (Larsson, 1993). While the case survey allows researchers to re-use the experiences found in the case studies (Jurisch et al., 2013), the focus in applying the method is the characteristics of the case rather than the analysis and conclusions made by original authors (Yin & Heald, 1975).

To select the case studies of RPA in the public sector, we performed a literature search in the Google Scholar database in April 2021. The reason to use Google Scholar is two folds (1) Google Scholar is becoming the first choice for research discovery, which offers an easy access to a large number of contents in which some may not be included in commercial databases. (2) Google Scholar is suitable as a source of data for scientific research which provides citations counts that are broader than other databases (Halevi et al., 2017). The search string was "Robotic Process Automation" AND "case study" AND "public sector." We filtered the search with the papers published after 2015 because the first RPA case was reported by Lacity et al. (2015). On June 20, we did a final search of the literature to see if there are new publications of RPA in public sector. Following Larsson's (1993) suggestions for eliminating the search specific biases, different publication types (e.g., research publications and conference papers) in a variety of journals and conference proceedings (e.g., Communications of the Association for Information Systems, MIS Quarterly, International Conference on Information Systems) were included in the search. The final hits included 220 publications. The inclusion criteria were: (1) the article is available in full text in English; (2) the article is a case study; (3) the article provides a qualitative evaluation of RPA; and (4) the article presents examples of RPA adoption and implementation in the public sector. Exclusion criteria were: (1) the article does not focus mainly on RPA; (2) the article is a business paper or marketing material written by a software vendor or a consulting firm; (3) the article is not a case study and (4) the case is not conducted in public organizations. These publications were reviewed in three subsequent rounds. First, the titles, then the abstracts and lastly, the main contents of the publications were read. A total of 8 case studies were included in the analysis, see Appendix. The selected case studies were analyzed according to their focal level of analysis (practice), unit of analysis (routines), and key findings (Swanson, 2019). Furthermore, the case studies were critically examined according to the four change modes (ibid.), namely design, diffusion, execution, and shift.

We performed thematic analysis (Braun & Clarke, 2006) with a theory-driven approach to aggregate the key insights from the 8 cases. Using Swanson's technology as routine capability as the theoretical lens, we applied the six-step model described by Braun and Clarke (2006) in the data analysis process: (1) Familiarization with data. The cases were read multiple times to become familiar with and to have a clear understanding. (2) Generating initial codes. The initial codes were generated based on process flows and practices described in the case studies. This step lasted until no new codes were identified. (3) Searching for themes among codes. As we applied a theory-driven approach, the themes were identified as "practices" and "routines" through the lens of routine capability. (4) Reviewing themes. The themes were reviewed to ensure that they were supported by the coded data. (5) Defining and naming themes. We refined and described our findings. (6) Producing the final report. We generated the scheme of analyzed cases (see Appendix). To ensure the reliability and validity of the results, the team performed the steps jointly. The consensus and agreements in interpretations were achieved through discussions and working meetings within the team.

4 Results

The case studies included in the analysis focus on the public organizations' RPA implementations. Though RPA adoption around the world is rapidly growing, our case sample covers 6 studies from Nordic countries (four from Sweden, one from Norway, and one from Finland) that studied RPA use in government agencies and organizations and 2 studies of RPA adoption in the public universities in India and Australia. We analyzed these case studies from the perspective of technology as routine capability. We identified that the current applications of RPA in the public sector advance citizen services (i.e., social assistance practices) and public organizations' internal services (i.e., HR and payroll practices). Moreover, we also find the changes caused by RPA implementation in public organizations.

4.1 How RPA Advance Business Process Management Practices

The analysis reveals that RPA as routine capability advances three contexts of practices in public organizations—individual, organizational, and social.

4.1.1 RPA Advancing Practices at the Individual Level

Advance Human Practice by Focusing on Value-Added Work and Developing New Skills

The results show that by being freed up from mundane work, public employees have more time to focus on the work requiring their analytical skills (Denagama Vitharanage et al., 2020; Dias et al., 2019; Ranerup & Henriksen, 2020). This shift in employees' focus advances human practice in two ways: First, it enhances the employees' job satisfaction and the feeling of self-fulfillment. As reported by Denagama Vitharanage et al. (2020), the respondents in their case study clearly explained the importance of the sense of fulfillment.

"Staff is actually using their actual analytical and problem-solving skills... To me, that's more interesting work as well from a job satisfaction than just sitting entering data... certainly, from a job satisfaction, it's going to be, in time."

Second, the employees can perform and participate in more value-adding work, which requires personal interaction and decision-making (Denagama Vitharanage et al., 2020; Ranerup & Henriksen, 2020). For example, Ranerup and Henriksen (2020) show that RPA as routine capability advanced human practice by providing the time and space to civil servants to focus on the critical elements of social assistance applications made by the citizens. These include reviewing the activity plans and providing expert opinions which require more time and professional experience in social assistance decision-making practices.

Advance Machine Practice Through Routine Capabilities of RPA

RPA provides its routine capabilities with its implementation. The new routines performed by RPA replace the human routines and build the machine practices. Our review shows that the examples of such machine routines include: in a payroll practice, processing various types of paper forms (used for handling sessional appointments, paper appointments, resignations, invitations to adjunct professors), validating and entering data (Denagama Vitharanage et al., 2020); in a social assistance decision-making practice, checking the application requirements, handling the applications with algorithmic evaluations, making decisions (Ranerup & Henriksen, 2020). Furthermore, RPA as routine capability advances the machine practices through interacting with different information systems. This interaction can be with the internal information systems as well as the ones external to the organization. Ranerup and Henriksen (2020) provide an example of external interactions through which RPA communicates information between different platforms, accessing the national platform for information in decision-making practices. Additionally, Goday-Verdaguer et al. (2020) report that RPA as routine capability could better improve machine practices if the process to be automated is well defined. According to the results of our analysis, advancing machine practices also includes the creation of combined routines where other software-enabled routines such as the ones associated with process mining practices.

4.1.2 Advance Organizational Practices

RPA as routine capability advances the organizational practices mainly through reducing manual work and improving efficiency (Denagama Vitharanage et al., 2020; Dias et al., 2019; Goday-Verdaguer et al., 2020; Lindgren, 2020; Ranerup, 2020; Ranerup & Henriksen, 2020).

The analysis indicates that RPA advanced various practices, as shown in the case studies in our sample. The advanced practices include human resources (HR) practices (Dias et al., 2019), payroll practices (Denagama Vitharanage et al., 2020), and decision-making practices (Ranerup, 2020; Ranerup & Henriksen, 2019; Ranerup & Henriksen, 2020).

With regards to payroll practices, the results show that implementation of RPA advances the payroll practices through the reduction in time (i.e., process cycle time, task handling time, waiting time, turnaround time), decrease in the errors (i.e., in data entry), improved correctness and timing of the customer assistance services (Denagama Vitharanage et al., 2020). Further, operational reliability in payroll practices is achieved through the continuity of services (through RPA's capability to work continually) in times of unexpected situations. These advancements in the payroll practice, in turn, have successive positive effects on managerial practices, such as ensuring the application of business rules, fulfilling the compliance requirements, and optimizing the use of human resources (Denagama Vitharanage et al., 2020).

At a more strategic level, RPA as routine capability provides improvements regarding transparency and visibility (Denagama Vitharanage et al., 2020). This capability helps organizations better understand processes and identify the responsibilities of stakeholders of the processes.

In addition to these, RPA as a routine capability improves the competitive advantage of the organizations as an early adopter of emerging technological routine capabilities (Denagama Vitharanage et al., 2020).

Our analysis indicates that RPA as routine capability advances the decision-making practices (Ranerup, 2020; Ranerup & Henriksen, 2020). Most significantly, legal certainty of the decisions on social assistance is increased through the routine capabilities of RPA (Ranerup & Henriksen, 2020). According to Ranerup and Henriksen (2020), this advancement will positively impact the uniformness and fairness of the decision-making practices of public organizations in the future.

4.1.3 Advance Social Practice

The analysis shows that RPA supports digital government initiatives in the public sector. The cases from the public sector reveal that government organizations implemented RPA in alignment with public sector digitalization agendas which guides the society in digital transformation (Dias et al., 2019; Ranerup, 2020; Ranerup & Henriksen, 2019; Ranerup & Henriksen, 2020). The results also show that RPA as routine capability advance social practice by improving decision-making practices (Ranerup & Henriksen, 2020). Improving legal certainty of the social assistance decisions (Ranerup & Henriksen, 2020) has positive impacts on both society and individual and thus advances social level practices.

4.2 Advancing Business Management Practices

In this section, we present the results regarding what changes RPA as routine capability through the four modes of change in advancing the practices: (1) design, (2) execution, (3) diffusion, and (4) shift.

Change by design refers to advancing a practice by creating a new device and routine or improving the current devices and routines to gain new capabilities (Swanson, 2019). The results indicate RPA as routine capability advances the practices through change by design when certain conditions are met. This means that technology as routine capability develops when the constitutive devices provide affordances. For example, in order to automate the decision-making process, e-applications are necessary pre-conditions (Ranerup, 2020). Goday-Verdaguer et al. (2020) show that streamlining the processes is a prerequisite for RPA implementations.

Change by execution is more about learning through practices which entails continuous improvement and improvisation (Swanson, 2019). RPA as routine capability has two aspects with regards to change by execution. The first aspect is related to the learning process of software robots (Güner et al., 2020). In that, the software robots, as agents of automation, can learn by observing humans executing complex processes (van der Aalst et al., 2018). More frequently (with standard RPA), the software robots are programmed to perform the task previously done by humans. On the other hand, software robots with AI capabilities can be trained with data and thus adapt to different situations. Examples include the practices of decision-making, which are built on the routine capabilities of AI-enhanced RPA (Ranerup & Henriksen, 2019, 2020).

Another aspect of change by execution is associated with routines that require proficiency. That is changes brought by RPA advance the practices through supporting the skills development and utilizing the employees' experiences. Our analysis shows that the execution of algorithmic decision-making changes the social assistance practices and the focus of the caseworkers toward critical tasks. For example, Dias et al. (2019) show that RPA adoption enhanced cognitive reasoning and advanced human practice in the case of organization's employees.

Change by diffusion refers to the diffusion of new routine capabilities to a broader population for advancing a practice. The results show that RPA as routine capability is diffusing across public organizations.

Ranerup (2020) clearly illustrates how RPA implementation in Trelleborg Municipality has been a pioneer in the automation of social assistance practices and disseminated by the authorities (Swedish Association of Local Authorities and Regions- SALAR) as a new working model for social assistance practices. Through several dissemination activities (i.e., workshops and visits to Trelleborg Municipality), RPA as routine capability has been promoted by the authorities to advance the social assistance practices nationwide. However, Lindgren (2020) claims that RPA implementation in municipalities is oversimplified considering the challenges in changing the routines of administrative work and emphasizes the necessity of redesign of the processes in order to fully benefit from the capabilities of RPA in advancing practices.

Change by shift in technology as routine capability is regarded as the emergence of new practices through adapting and recreating routines to develop new capabilities for advancing human practices (Swanson, 2019). According to Swanson (2019), practices do not persist in the same frequency or intensity as they are, nor do the associated routines and capabilities. The results show that RPA as routine capability has changed the routine capabilities. Examples include payroll practices (Denagama Vitharanage et al., 2020), and decision-making practices (Ranerup, 2020; Ranerup & Henriksen, 2020). These changes in practices have broader impacts on the shifts among associated (other) practices. For example, in decision-making practices, RPA as routine capability engages not only the employees that perform the new routine but also the population of municipal workers with regards to the new way of designing and performing routines (Ranerup, 2020). Similarly, Lindgren (2020) predicts that the change in administrative routines will create a shift in the work-life conditions and content.

5 Discussion and Conclusion

In this chapter, we aimed to present and analyze the use of RPA in public organizations through the theoretical lens of technology as routine capability. The study shows that a few early adopters have implemented RPA in the public sector, where the Nordic countries are in the most prominent positions.

The case survey shows that RPA creates new "machine" routines and becomes integral to humans' new routines. RPA as routine capability does advance public service practices at individual, organizational, and social levels though we only have evidence from 8 cases found in literature. At the individual level, RPA as routine capability supported the employees to increase their job satisfaction and sense of fulfillment. Organizational level advancements in public organizations are mostly associated with the public service quality improvements, i.e., in social assistance. The advancements are found in public organizations such as municipalities and universities. At the social level, the study shows that RPA as routine capability advances social practice by the application of central governments' digitalization strategies at the local levels (i.e., in municipalities). Another social level advancement is the improvement in legal certainty of the decision-making practices which has a positive effect on the society and the individual. The evidence also indicates that changes of RPA are intertwined in the four modes in the public organizations: design, execution, diffusion, and shift. The use of RPA triggers new e-application designs for improving public services and requires new skills developments for the public workers to focus more on the critical tasks. The advancements in one practice in the public organization are seen to be diffused to other public services. The changes also indicate the public services to shift from routines to design and implement new RPA-enabled practices.

In this case survey, we also find that RPA as routine capability advanced the decision-making practices through increasing transparency (Ranerup & Henriksen, 2019). Furthermore, uniformness and fairness in discretion practices can be achieved, as revealed by the reviewed case studies (Ranerup, 2020; Ranerup & Henriksen, 2020). However, Smith et al. (2010) argue that human decision-making, which is necessary in certain complex conditions, is restrained by automating discretionary practices. While RPA as routine capability advances human and machine practices, the new routine capabilities provided by RPA may further cause a decrease in employment in the public sector. Previous research (i.e., Borry & Getha-Taylor, 2019; Eikebrokk & Olsen, 2020; Söderström et al., 2021), indicates that there is an ongoing debate around employment lay-offs due to the automation of public services, especially more cognitive functions and AI technologies extend RPA to solve more complex tasks based on unstructured data or images and videos. Although managing internal communication and redeployment are essential for public workers to approach RPA positively and develop new skills for more intellectual work, how can these workers' practices be advanced if RPA completely replaces their "old" practice? The resistance to new technology should be predicted and addressed by the management to eliminate the fear of job loss. This also challenges the core assumption of technology as routine capability "*human seek to advance their practices*" (Swanson, 2019, p.1011). No concrete evidence for this assumption is proposed from the cases. It is also expected that the employees who are comfortable with their job positions may be reluctant to such change, which may result in the failure of RPA implementation in public organizations.

Technology as routine capability is a new analysis framework, it has helped us focus more on the practice side of public services improvement through RPA and provides good support for analyzing changes at different levels. There are other studies that cite Swanson (2019) such as Mendling et al. (2020) and Pentland et al. (2020). Although the framework for technology as routine capability is not used directly, these studies acknowledge the intertwinement of routines, BPM technologies and process changes (Mendling et al., 2020) by referring to Swanson (2019). Swanson's framework helped us identify changes in practice that we do not think we would have identified easily. However, Swanson (2019) did not provide clear guidelines regarding how to analyze each case through "zoom in" and "zoom out" conduct, as well as how to aggregate and disaggregate the insights from multiple cases. This is a method issue. We argue that to increase this perspective's applicability, more guidelines for conducting case studies and analyzing data are necessary.

A limitation of this study concerns our understanding of routines, practices, and their conceptual meanings that can be theoretically understood from "processes and services" in the context of public sectors. We interpreted the changes in "processes and services" as equal to changes in routines and practices when analyzing the eight cases. The RPA implementations triggered immediate changes in "processes and services routines," but we should acknowledge that new routines and the capability provided by the new routines in advancing human practices cannot be achieved simultaneously (May et al., 2009). Longitudinal studies are needed to explore how

RPA as routine capability enacts new practices and service routines in the public sector.

This study contributes to a deeper understanding of the development and use of RPA in public organizations and how it generates new routine capability in advancing human and machine practices. Technology as routine capability provides a new perspective to gain a better knowledge of how RPA advances public service practices. The chapter presents the state-of-art knowledge that can be useful for continuing RPA research in public administration and the digitalization of public services.

We conclude that in this study, we confirm the fruitfulness and applicability of technology as routine capability as the analytical framework to accumulate insights from different cases. However, as RPA is in the stage of early adoption within the public sector, the available cases did not provide profound insights for the analysis. We expect future research on more advanced applications of RPA to produce more novel insights on routine capability changes within public organizations and to contribute with more knowledge about digital transformation of public organizations.

Appendix: Case Studies Included in the Analysis

No	Author, Year, Public sector, Country	Focal level of analysis/Practice	Unit of analysis/ Routines	Key findings
1	Dias et al. (2019), Finnish government shared services center (FinServ), Finland	Digital transformation practice (HR and Finance processes digitalization) Add more which organization, from which country, discuss the differences and similarities	*Knowledge workers' routine in human resources HR & finance processes* This case did not describe any detailed information of specific routines or processes. The case studied the business processes in the Finnish government shared service Centre (FinServ). The goal was to get about 40% efficiency to the processes after RPA implementation	Advance knowledge workers' practice to focus on more value-added tasks in HR & Finance Management found that knowledge workers are capable of doing more intellectual tasks than they thought. However, how the RPA fuses with the knowledge work routine and provides capabilities for advancing knowledge workers' practice needs proper management and coordination. Human attention was still

(continued)

No	Author, Year, Public sector, Country	Focal level of analysis/Practice	Unit of analysis/ Routines	Key findings
				needed to handle highly cognitive problem-solving tasks and develop new knowledge for continuous business development
2	Patil et al. (2019), The Savitribai Phule Pune University, India	Examination result analysis practice in a university	*Routines in data transfer for examination assessments* Old routine: Transferring the examination results in PDF to excel file manually New routine: Transferring the examination results in PDF to excel automatically without human intervention before the proper analysis for each subject	Advance examination assessment practice by reducing the completion time and errors of data transfer from PDF to Excel
3	Ranerup and Henriksen (2019), Municipality, Sweden	Digitalization and automation of decision-making practices in social services	*Routines in social assistance decision-makings in public organizations* The case study (Trelleborg Municipality, Sweden) doesn't provide information of practices before and after the implementation of RPA New Routine generates capability: Beginning in the Spring of 2017, social assistance decisions were increasingly managed by the RPA. By August 2017, RPA, assisted by caseworkers, handled approximately 70% of the applications. Of these applications, RPA made 41% of the decisions	Though RPA implementation provides increased accountability, enhanced efficiency, and cost reduction for social services, the practice of decision-making involves human judgment, exceptions, along the values that shape public sector services. The confidence in human worker's practice is already considered as advanced in terms of professionalism

(continued)

No	Author, Year, Public sector, Country	Focal level of analysis/Practice	Unit of analysis/ Routines	Key findings
			and handled the actual social assistance payments. If applications were rejected, caseworkers from the Labor Market Agency handled the decisions manually	
4	Denagama Vitharanage et al. (2020), University, Queensland University of Technology (QUT), Australia	Payroll practices (including payroll process and its sub-processes such as sessional appointments, paper appointments, resignations, invitation to adjunct professor)	*Routines in data processing and data entry* Old routine: Entering the paper forms to the system manually by the payroll team New routine: The case provides that RPA implemented into the payroll process. Although there is no further information on practices before and after the implementation of RPA, the case provides the benefits gained by the implementation	Advance payroll practices by improving time efficiency, improvement in the accuracy, improving process transparency, visibility and understanding, decrease in the compliance risk (e.g., protecting the university from potential litigation, being compliant with regards to contracts) Advance human practice and skills through supporting the employees on the development of new knowledge on automation and engagement in more value-adding work Advance organizational practices by enforcing organizational policies, increasing customer satisfaction and operational continuity
5	Lindgren (2020), Municipality, Sweden	Practice of case handling in social work and administrative processes in local governments	Old and new routines aren't described	The authors expect that the automation of the administrative routines will have challenging impacts on the employees' work life and the content of their work

(continued)

No	Author, Year, Public sector, Country	Focal level of analysis/Practice	Unit of analysis/ Routines	Key findings
6	Ranerup (2020), Municipality, Sweden	Changes related to RPA in decisions on social assistance	*Automation of decision-making practices in social work (case handling in social assistance) and dissemination of RPA at national level* Old and/or new routines aren't described	What changes RPA as routine capability. Four change modes revealed by the two cases are as follows. 1. Change by design in decision-making routines 2. Change by execution in learning and adapting in different situations 3. Change by diffusion in spreading the RPA as routine capability 4 Change by shift through re-creating and performing the decision-making routines in a broader population
7	Ranerup and Henriksen (2020), Municipality, Sweden	Practice of social assistance application and decision-making Practice of discretion in social assistance decision-making	*Current routine: Routines in social assistance decision-makings in local government (the separate routine for decision)* Social assistance application is a multi-level process including several steps in each level. Following the citizens' applications for social assistance, RPA routines are performed. The steps in routine are as follows: 1. Routines preliminary to the social assistance decision: (a) RPA bot is programmed as a caseworker (b) A citizen's application form is	RPA advanced decision-making practice through: The improvements in objectivity, accountability and efficiency Decrease in the errors and cost Time savings for the caseworkers However, some of the caseworkers showed resistance to the transfer of responsibility from caseworker to RPA, claiming that responsibility transfer pulled down their professionalism. Although RPA replaces manual automated decision-making practice, the expertise of the

(continued)

No	Author, Year, Public sector, Country	Focal level of analysis/Practice	Unit of analysis/ Routines	Key findings
			copied and entered into the system (manual routine) (c) The robot logs into the case management system ProCapita as a caseworker, copies the information from the form, transfers the information to an Excel document to initiate checking with the social insurance board or another agency. (d) the robot checks if citizens have an operational activity plan 2. RPA makes the decisions for the applications (one-third of the decisions are made by RPA). Still, the robot and human caseworker make the final decision jointly in many cases 3. complex decisions are handled by human caseworker and RPA jointly	caseworkers is still needed (particularly in complex cases/ applications) RPA as routine capability needs to be improved for making decisions in complex situations
8	Goday-Verdaguer et al. (2020), Municipality, Norway	Human resources (HR) practices	Current routines in the hiring process (based on the process analysis) After the position is announced publicly, a new case is created in the recruitment system Preparation: The job analysis is conducted by the hiring unit The job analysis is approved The job	RPA is expected to advance organizational efficiency through improving HR practices

(continued)

No	Author, Year, Public sector, Country	Focal level of analysis/Practice	Unit of analysis/ Routines	Key findings
			advertisement text is created A case is opened in the HR system, and the job advertisement is published A list of applications is created Selection: Selection is out of the scope of the HR department. After the decision is made, HR stores the final decision consisting of a ranking of candidates, provides an offer for the most suitable candidate and negotiates the contract Finalization: After the candidate accepted the position and signed the contract Several systems are accessed, and ordering of equipment is performed	

References

Aguirre, S., & Rodriguez, A. (2017). Automation of a business process using robotic process automation (RPA): A case study. In J. Figueroa-García, E. López-Santana, J. Villa-Ramírez, & R. Ferro-Escobar (Eds.), *WEA 2017* (Vol. 742, pp. 65–71). CCIS.

Borry, E. L., & Getha-Taylor, H. (2019). Automation in the public sector: Efficiency at the expense of equity? *Public Integrity, 21*(1), 6–21.

Braun, V., & Clarke, V. (2006). Using thematic analysis in psychology. *Qualitative Research Journal, 3*(2), 77–101.

Cewe, C., Koch, D., & Mertens, R. (2018). In E. Teniente & M. Weidlich (Eds.), *BPM 2017 minimal effort requirements engineering for robotic process automation with test driven development and screen recording* (Vol. 308, pp. 642–648). LNBIP.

Denagama Vitharanage, I. M., Bandara, W., Syed, R., & Toman, D. (2020). An empirically supported conceptualisation of robotic process automation (RPA) benefits. In *Paper presented at the 28th European Conference on Information Systems (ECIS2020), Marrakesh, Morocco, June 15–17.*

Dias, M., Pan, S., & Tim, Y. (2019). Knowledge embodiment of human and machine interactions: Robotic-process-automation at the Finland government. In *Presented at the 27th European Conference on Information Systems (ECIS2019), Stockholm Uppsala, Sweden, June 8–14.*

Eikebrokk, T. R., & Olsen, D. H. (2020). Robotic process automation and consequences for knowledge workers; a mixed-method study. In M. Hattingh, M. Matthee, H. Smuts, I. Pappas, Y. K. Dwivedi, & M. Mäntymäki (Eds.), *Responsible design, implementation and use of information and communication technology* (Vol. 12066, pp. 114–125). Springer. https://doi.org/10.1007/978-3-030-44999-5_10

Feldman, M. S., & Pentland, B. T. (2003). Reconceptualising organizational routines as a source of flexibility and change. *Administrative Science Quarterly, 48*, 94–118.

Goday-Verdaguer, A., Mannhardt, F., & Torvatn, H. Y. (2020). Mining e-mail conversations to enrich event logs: An exploratory case study of a hiring process in a Norwegian municipality. In *Norsk konferanse for organisasjoners bruk at IT* (Vol. 28, No 1).

Güner, E. O., Han, S., & Juell-Skiels, G. (2020). Robotic process automation as routine capability: A literature review. In *Paper presented at the 28th European Conference on Information Systems (ECIS), an online AIS conference, June 15–17.*

Halevi, G., Moed, H., & Bar-Ilan, J. (2017). Suitability of Google scholar as a source of scientific information and as a source of data for scientific evaluation—Review of the literature. *Journal of Informetrics, 11*(3), 823–834. https://doi.org/10.1016/j.joi.2017.06.005

Houy, C., Hamberg, M., & Fettke, P. (2019). Robotic process automation in public administrations. In *Digitalisierung von Staat und Verwaltung. Lecture notes in informatics Vol 291, Köllen, Bonn.*

Jimenez-Ramirez, A., Reijers, H. A., Barba, I., & Del Valle, C. (2019). A method to improve the early stages of the robotic process automation lifecycle. In P. Giorgini & B. Weber (Eds.), *Advanced information systems engineering. CAiSE 2019. Lecture notes in computer science* (Vol. 11483, pp. 446–461). Springer.

Jurisch, M., Wolf, P., & Krcmar, H. (2013). Using the case survey method for synthesizing case study evidence in information systems. In *Presented at the Americas Conference on Information Systems (AMCIS2013) Chicago, Illinois, USA on August 15–17.*

Lacity, M. C., Willcocks, L. P., & Craig, A. (2015). *Robotic process automation at Telefónica O2. The outsourcing unit working research paper series (15/02).* The London School of Economics and Political Science.

Larsson, R. (1993). Case survey methodology: Quantitative analysis of patterns across case studies. *Academy of Management Journal, 36*(6), 1515–1546.

Lindgren, I. (2020). *Exploring the use of robotic process automation in local government. EGOV-CeDEM-ePart 2020* (pp. 249–258).

May, C. R., Mair, F., Finch, T., et al. (2009). Development of a theory of implementation and integration: Normalisation process theory. *Implementation Science, 4*(29), 4–29.

Mendling, J., Pentland, B. T., & Recker, J. (2020). Building a complementary agenda for business process management and digital innovation. *European Journal of Information Systems, 29*(3), 208–219.

Patil, S., Mane, V., & Patil, P. (2019). Social innovation in education system by using robotic process automation (RPA). *International Journal of Innovative Technology and Exploring Engineering, 8*(11), 3757–3760.

Pentland, B. T., & Feldman, M. S. (2008). Designing routines: On the folly of designing artifacts, while hoping for patterns of action. *Information and Organization, 18*, 235–250.

Pentland, B. T., Recker, J., Wolf, J. R., & Wyner, G. (2020). Bringing context inside process research with digital trace data. *Journal of the Association for Information Systems, 21*(5), 1214–1236.

Penttinen, E., Kasslin, H., & Asatiani, A. (2018). How to choose between robotic process automation and back-end system automation? In *Paper presented at the 28th European Conference on Information Systems (ECIS 2018), Portsmouth, UK, 23-28 June.*

Ranerup, A. (2020). Translating robotic process automation in social work: Aspirational changes and the role of technology. In *Presented at the 11th Scandinavian Conference on Information Systems (SCIS 2020) Sundsvall, Sweden, 9–11 August.*

Ranerup, A., & Henriksen, H. Z. (2019). Value positions viewed through the lens of automated decision-making: The case of social services. *Government Information Quarterly, 36*(4), 101377.

Ranerup, A., & Henriksen, H. Z. (2020). Digital discretion: Unpacking human and technological Agency in Automated Decision Making in Sweden's social services. *Social Science Computer Review*. https://doi.org/10.1177/0894439320980434

Ratia, M., Myllärniemi, J., & Helander, N. (2018). Robotic process automation—Creating value by digitalising work in the private healthcare. In *Proceedings of the 22Nd international academic Mindtrek conference, Mindtrek '18*. ACM.

Schou, J., & Morten, H. (2018). *Digitalization and public sector transformations*. Palgrave Macmillan.

Smith, M. L., Noorman, M. E., & Martin, A. K. (2010). Automating the public sector and organizing accountabilities. *Communications of the Association for Information Systems, 26* (1), Article 1. https://doi.org/10.17705/1CAIS.02601

Söderström, F., Johansson, B., & Toll, D. (2021). Automation as migration?—Identifying factors influencing adoption of RPA in local government. In *Paper presented at the 29th European Conference on Information Systems (ECIS 2021), Marakesh, Morocco, 14–16 June.*

Suri, V. K., Elia, M., & van Hillegersberg, J. (2017). Software bots – The next frontier for shared services and functional excellence. In I. Oshri, J. Kotlarsky, & L. Willcocks (Eds.), *Global sourcing of digital services: Micro and macro perspectives. Global Sourcing 2017. Lecture notes in business information processing* (Vol. 306, pp. 81–94). Springer.

Swanson, E. B. (2019). Technology as routine capability. *MIS Quarterly, 43*(3), 1007–1024.

van de Weerd, I., Nieuwenhuijs, B., Bex, F., & Beerepoot, I. (2021). Using AI to augment RPA: A conceptual framework. In *Paper presented at the 29th European Conference on Information Systems (ECIS 2021), Marakesh, Morocco, 14–16 June.*

van der Aalst, W. M. P., Bichler, M., & Heinzl, A. (2018). Robotic process automation. *Business & Information Systems Engineering, 60(4)*, 269–272. https://doi.org/10.1007/s12599-018-0542-4

Willcocks, L., Lacity, M., & Craig, A. (2015). *The IT function and robotic process automation. The outsourcing unit working research paper series*. London School of Economics and Political Science.

Yin, R. K., & Heald, K. A. (1975). Using the case survey method to analyse policy studies. *Administrative Science Quarterly, 20*, 371–381.

Organizing for Robotic Process Automation in Local Government: Observations from Two Case Studies of Robotic Process Automation Implementation in Swedish Municipalities

Ida Lindgren, Maria Åkesson, Michel Thomsen, and Daniel Toll

1 Introduction

This chapter reports on ongoing research into robotic process automation (RPA) in Swedish municipalities. Drawing on concepts from literature on RPA and the discourse on what benefits RPA can bring to public administration, we illustrate how the implementation of process automation with RPA is organized in two Swedish municipalities.

Currently, public organizations in countries worldwide are increasingly digitalizing and automating work. This trend may have profound implications for governments, local authorities, and citizens, as it affects how public organizations are organized. It also affects the time, place, and nature of these organizations' interactions with citizens (Lindgren et al., 2019). In Sweden, digital transformation is pushed by multiple actors in society, such as the government, governmental agencies, and the Swedish Association of Local Authorities and Regions (SALAR). SALAR is an employer's organization that represents and advocates for local government in Sweden. The Swedish national digitalization strategy (Government Offices of Sweden, 2017) underpins the parliament's goal that Sweden is to be world leading in digitalization. This overall goal is divided into sub-goals, one being *digitalization management*. This goal refers to that purposeful and relevant digitalization can drive quality management and streamlining of public services in Sweden. On a similar note, SALAR has published policy documents highlighting various forms of automation technologies as a way of accelerating the

I. Lindgren (✉) · D. Toll
Department of Management and Engineering, Linköping University, Linköping, Sweden
e-mail: ida.lindgren@liu.se; daniel.toll@liu.se

M. Åkesson · M. Thomsen
School of Information Technology, Halmstad University, Halmstad, Sweden
e-mail: maria.akesson@hh.se; michel.thomsen@hh.se

G. Juell-Skielse et al. (eds.), *Service Automation in the Public Sector*, Progress in IS,
https://doi.org/10.1007/978-3-030-92644-1_10

transformation of regional and local government organizations toward becoming more efficient and cost-effective (SALAR, 2018a, b). One of the automation technologies promoted by SALAR is Robotic Process Automation (RPA).

RPA is software designed to perform administrative tasks in existing IT systems' user interfaces, typically in the same way an employee would use the systems. To operate in a system, the software or "robot" is assigned a user ID and passwords required for the job. Simply put, these software robots imitate human worker's ways of performing rule-based administrative tasks (see e.g. Hallikainen et al., 2018; Lacity & Willcocks, 2016; Willcocks & Lacity, 2016). Currently, RPA is implemented in several Swedish municipalities. However, there are concerns about what public values RPA can create (Lacity & Willcocks, 2017) and signs of unrealistic expectations from the implementations (Trefler, 2018). More importantly, implementations of this type of software may have profound implications for public administration, professional discretion, work environment, and interactions with citizens (Åkesson & Thomsen, 2020). A key question for municipalities pursuing process automation with RPA is how to organize the implementation initiatives (Lindgren et al., 2021). It is therefore important to explore and understand how RPA initiatives are organized and implemented in public administration.

A general pattern can be seen in Sweden and beyond; the use of RPA is promoted by policymakers and private IT suppliers as a suitable technology for government organizations to interconnect systems and create more efficient workflows. Yet, there is limited practical experience and research on the implementation and outcomes of automation of public services using RPA and its associated risks (see, e.g., Ngwenyama et al., 2021). It remains an empirical question what organizational consequences are brought by increased automation of government operations using RPA. In this chapter, we use empirical examples to illustrate how RPA initiatives are organized and implemented in Swedish municipalities. The aim of the chapter is to illustrate general patterns of how two Swedish municipalities organize the implementation of RPA, and associated challenges. We particularly focus on new roles and structures that have been introduced to support RPA implementation in the organizations. We contribute with empirical illustrations that show how general policies on process automation, promoted by policymakers (e.g., SALAR), are influencing the organization of IT and work in local government practice. This, in turn, illustrates challenges that affect the implementation of RPA in local government and point to a set of observations that require further research.

The chapter is organized as follows: First, we provide the background of our research followed by our research approach. We then continue to present empirical illustrations from two municipalities. Finally, we discuss these illustrations and summarize our main observations in the conclusions section.

2 Background

Swedish local government is made up of 290 separate municipalities. The municipalities are self-governed to a large extent, although adhering to the Swedish Local Government Act. This act stipulates that all citizens, no matter what municipality they live in, have the right to obtain municipal services. These services include, e.g., social services, care for children and the elderly, public schools, emergency services, environmental services, and building permissions.

An integral part of municipal service delivery is administrative work related to information handling and decision-making for service eligibility (i.e., *case handling*). In most Swedish municipalities, case handling is still very much a manual labor for employees (SALAR, 2018b). Oftentimes, a large part of the administration involves manual copying and pasting information from one system or form to another system. This manual work is associated with high costs for human laborers, as well as a risk of human error. For these reasons, automation of administrative work is currently seen and marketed as a way of reducing costs and reducing errors in the information handling process (SALAR, 2018a, b; SOU, 2014, 2016). Each municipality decides how, and to what degree, its operations and services can and should be digitalized and automated. However, although self-governed, the municipalities are under pressure to automate administrative work from more "soft" governance initiatives, such as the policy documents mentioned previously.

A key question for municipalities pursuing process automation with RPA is how to organize the initiatives (Lindgren et al., 2021). At the core is the relationship between RPA and traditional IT initiatives (see e.g. Osmundsen et al., 2019) and whether it is appropriate to treat RPA as any other IT initiative, or if it is more appropriate to drive the RPA initiatives as a specific type of digitalization. Another issue regards centralization versus de-centralization, where previous studies point to different recommendations. One way of organizing for RPA is to use a centralized, top-down, model where a center of excellence or similar takes the lead and responsibility of RPA initiatives (Schmitz et al., 2019). A different approach is to use a de-centralized, bottom-up, approach where departments autonomously develop RPA (Willcocks et al., 2015). A third approach is to use a middle-out approach where the RPA initiatives are driven by departments with the support from a center of excellence, as suggested by Noppen et al. (2020). The three models, centralized, de-centralized, or a combination of both, are all associated with advantages and disadvantages. In a study of RPA initiatives in three organizations, Osmundsen et al. (2019) found that there are advantages with organizing and managing RPA initiatives in the departments, i.e. outside of the central IT function. This model led to the departments building enthusiasm for digitalization and local ownership. The model was however also accompanied by a lack of control and a lack of end-to-end process view. A study of the third approach, i.e., to let departments drive the development with the help of an RPA center of excellence, showed problems of balancing centralization and local needs (Asatiani et al., 2019). All in all, there are several difficulties related to the organizing of RPA, and there is no best practice solution. It is clear, however, that RPA initiatives require detailed domain knowledge of the departments and their processes, to be combined with the kind of IT competence typically found in the IT department (Osmundsen et al., 2019; Willcocks et al., 2015).

3 Research Approach

This chapter builds on empirical data generated through two ongoing case studies[1] of RPA development and implementation at *Municipality East*, and *Municipality West*. Our work is based on a qualitative and interpretive approach (Myers, 2009; Walsham, 1995), meaning that we are interested in how people interpret and understand RPA and automation of case handling processes and, in turn, how they organize their RPA implementation in terms of formal structures and roles.

The municipalities included in this study were chosen based on being similar in size (both are classified as being large-size municipalities) and having comparable conditions for RPA implementation. The organizations are anonymized and therefore called East and West in the accounts below.

The account of Municipality East is based on 21 semi-structured interviews conducted with 18 employees in the organization. The interviews were conducted between February 2020 and January 2021. First, members of the DigiGroup (see account below) were interviewed. Further interviewees were then identified through a snowball sampling technique (Patton, 1980). Each interview was conducted using Zoom or Teams, except one in-person interview. Each interview lasted for approx. 90 min. The interviewees were asked about their work at the municipality, their view on automation of work in general, their experiences of RPA as a technology, perceived consequences of RPA implementation, how IT and digitalization is organized in the organization, and other questions concerned with the municipality's work to further digitalization and automation of work.

The account of Municipality West is based on semi-structured interviews with 14 employees, and 6 group interviews with a total of 12 different employees. The interviews were conducted between February 2020 and March 2021. The initial interviews were conducted face-to-face. These interviews included the CIO of digitalization, the development leader, and key digitalization staff within the IT service department. The following interviews, which were group interviews, were conducted using Zoom or Teams. These interviews included the RPA group and operational staff who were directly involved in RPA development. All interviews lasted between 60 and 120 min. The interviewees were asked about digitalization in

[1]The respective case studies are part of two research projects funded by AFA Försäkring (eng. AFA Insurance). The case *Municipality East* is part of a project run by researchers at Linköping University (*see* Lindgren, 2020; Lindgren et al., 2021; Söderström et al., 2021; Toll et al., 2021). The case *Municipality West* is part of a project run by researchers at Halmstad University (*see* Åkesson and Thomsen, 2020).

general, the role of automation and RPA, about how RPA initiatives are organized, and their approach to RPA implementation. They were also asked about the implications of RPA implementation on work environment and public administration.

The empirical material was analyzed through an iterative interpretative analysis (Klein & Myers, 1999), in which empirical observations made in the interviews were continuously related to the research aim and existing theories. The empirical observations from each municipality were first analyzed separately and results from these analyses are published elsewhere (e.g., Åkesson & Thomsen, 2020; Lindgren et al., 2021; Toll et al., 2021). For the purposes of this chapter, the findings from each analysis were synthesized to enable a comparison between how the two municipalities have organized their efforts to implement RPA in their respective organization. This part of the analysis focused on new roles and structures that have been introduced in the two municipalities to support RPA implementation. This approach enabled a deeper understanding of different patterns of organizing for RPA and allowed for multiple perspectives and interpretations of the material (Myers, 2009). In this chapter, we present findings from our analysis as empirical illustrations of how RPA initiatives are organized and implemented in the two municipalities.

4 Empirical Illustrations

The empirical illustrations of the RPA initiatives in Municipality East and Municipality West describe how the RPA initiatives started, the actions taken to organize the initiatives, and the relation between the different actors involved. Both municipalities have set ambitious goals to adhere to the national digital agenda of Sweden, i.e., to take full advantage of digital technologies to enhance and streamline work. As mentioned previously, municipalities in Sweden are self-governed, thus the initiatives studied in this paper are two unrelated cases. Consequently, the initiatives to implement RPA in these organizations are two instances in a bigger landscape with multiple ongoing digitalization initiatives across Swedish municipalities.

4.1 Organizing for RPA in Municipality East

Since 2018, the municipality is undergoing extensive organizational changes with the purpose of leveraging digitalization across its various departments. These changes involve several interrelated actions, some of which affect the municipality's preparedness and ability to automate administrative processes:

Hiring a Director of Digital Transformation (Director DT). In 2018, a person with extensive experience in digitalization from a large private corporation was hired to lead the digital transformation of the municipality. The purpose of this role was to highlight and give priority to digitalization. The Director DT is positioned on the strategical level and directly subordinated the Municipal Director. As a first mission, he was asked to form a Digitalization Group.

Forming the Digitalization Group (DigiGroup). During a period of approx. 2 years, the Director DT formed the DigiGroup, made up of highly educated and experienced professionals, knowledgeable in IT, digitalization, and business development. This is a strategic group responsible for stimulating digital transformation. They have only a small budget that can be used for strategically important projects; digitalization is otherwise funded by the departments. In 2019, a business development specialist was added to the group, with responsibility to pave the way for automation solutions in the organization (here called the Automation Leader).

Reorganizing the IT department and implementing a new IT-governance model. Concurrently as the DigiGroup was formed, the IT department was incorporated in the municipality's Support Services. This involved a shift from being a stand-alone department, to having a more explicit focus on supporting and providing services to the rest of the municipality. Simultaneously, a new IT-governance model was rolled out, forming new roles and teams concerned with IT governance for each department. These teams include both IT personnel and operational staff, making IT a more integral part of the municipality's business processes.

RPA pilots and competence building. During 2019–2020, DigiGroup initiated and partly funded two RPA pilot projects. Both pilots were seen as strategically important to test RPA and build automation competence in the organization. However, both were perceived as very costly. After the pilots were completed, the responsibility for the implemented solutions was handed over to the departments. Simultaneously, the Automation Leader worked on multiple initiatives to advocate, inspire, and educate employees on the possibilities of process automation (incl. RPA). This was done by participating in, e.g., department meetings, organizing workshops with managers, and guiding method development. From this work, an additional RPA development project was initiated by a business developer in one of the departments. Figure 1 depicts the main stakeholder groups involved with different roles and responsibilities relating to RPA within the organization.

As can be seen above, much effort has been put into creating new structures for dealing with IT and digital transformation in the municipality, including automation. The Automation Leader has been working on multiple fronts to create awareness of the potential benefits of automation; nevertheless, the adoption of this technology in the organization has been slow. In June 2021, three RPA applications were in use. The slow development can partly be explained by the setup and responsibilities distributed across the involved actors. The DigiGroup's role is strictly strategic, but managers in the departments often expect them to be operational: "*the managers are stressed to manage their day-to-day business and have not yet incorporated digitalization as a part of regular business development, they expect someone else to digitalize for them*" (quote from interview with the Automation Leader in 2021). In addition, the IT department's focus lies on maintenance; from their perspective, RPA is perceived as an unstable technology. The IT personnel, therefore, advocates for more conventional solutions to achieve automation (e.g., integration of systems), creating mixed messages concerning RPA. As a result, DigiGroup is somewhat

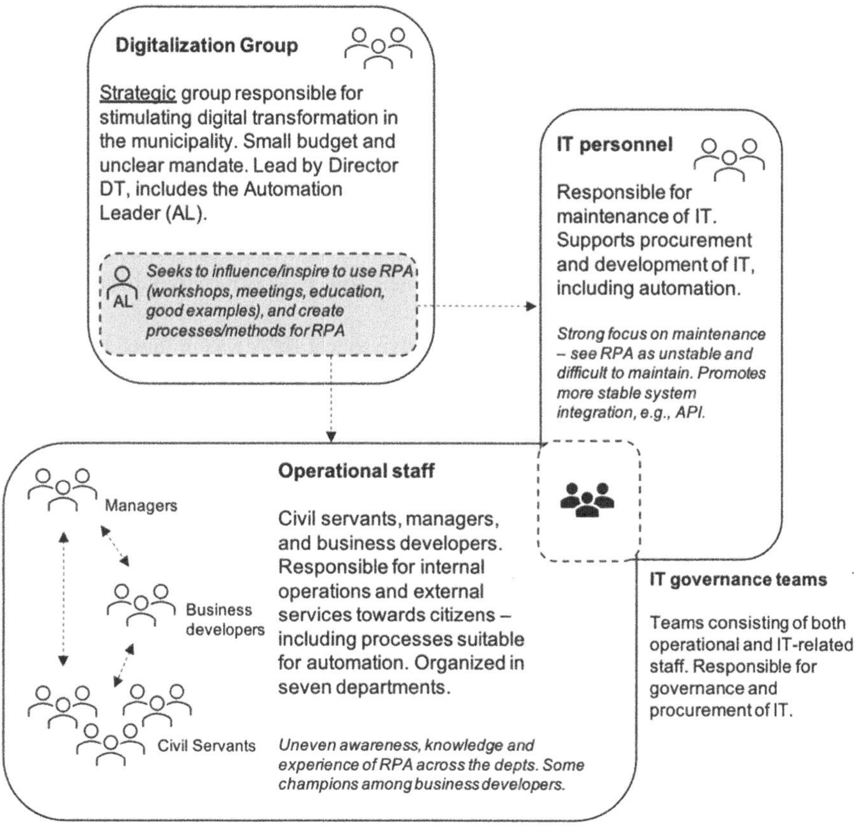

Fig. 1 The main stakeholder groups with their accompanied roles and functions in relation to RPA in Municipality East

struggling to find their role in an organization where those who could benefit from process automation are too busy and not knowledgeable enough about this technology (the operational staff). Moreover, those knowledgeable about automation and could help implement this technology are prioritizing stability and maintenance over new developments (the IT personnel). As a response, the Director DT and Automation Leader formed strategic and tactical groups in the beginning of 2021. These groups were formed to create a link between the Director DT and top managers (strategic group), and the Automation Leader and middle management (tactical group). At the time of the last interview (June 2021), these groups had been formed but were not yet operational. The goal for the continued development is to set a development process for automation in general (including RPA initiatives) and continue to educate and inspire employees in the departments to see the potential with process automation.

4.2 *Organizing for RPA in Municipality West*

Municipality West started to organize for automation with RPA in 2018 with the ambition to reduce manual administration, improve service quality for citizens, and, to become a more attractive employer. The municipality has organized several activities and roles to pursue this ambition:

CIO of Digitalization and a digitalization advisory group. In 2015, a CIO was hired, and a digitalization advisory group was formed. Both are positioned on the strategic level, within the City Council. Their role is to drive digitalization in the municipality.

Pilot initiative for RPA. The first initiative to test RPA was initiated by the CIO of Digitalization and conducted in 2018–2019. The ambition was to evaluate the potential of automation with RPA for the municipality. A consultant was involved, and the pilot involved the social care department in collaboration with department staff and IT service staff. As one of the outcomes of the evaluation of this initiative, it was decided to proceed with RPA initiatives on larger scale. A goal was set to have 30 RPA applications by the end of 2020.

Digital development team within the IT service department. After the first RPA pilot, a specific team was formed within the IT service department, assigned to stimulate digitalization initiatives. These initiatives can be e-services, RPA initiatives, and other new digital solutions. The group is managed by a team leader who is not an IT professional (but was part of the RPA pilot). This group is meant to be attentive to organizational needs for digitalization, as experienced and expressed by employees in the municipality.

RPA-group within the digital development team. In 2019, a specific group was formed to be responsible for RPA specifically. They engage with the departments, the IT service staff, and the consultants to stimulate, initiate, and realize RPA initiatives. The group also governs the RPA applications after implementation.

A municipality digitalization plan and fund. In 2019, a new digitalization plan for 2020–2023 was approved by the municipality board. This document presents the vision for the overall digitalization in the municipality and highlights automation as one of the most important means for cost-efficient, available, and high-quality services for citizens. To stimulate digitalization initiatives, a digitalization fund was established in 2019 from which the different departments within the municipality can apply for funding for development projects, including RPA implementations.

Reorganization of digitalization advisory group and RPA group. As of May 2021, the digitalization advisory board was replaced by a new department for the development and management of digitalization. The CIO of the new department is positioned directly under the municipal chief executive. The first assignment for the new CIO was to determine and suggest a new organization for digitalization within the municipality to be implemented by late 2021. Similarly, from September 2021, the RPA-group was reconstructed to an automation group with a wider scope of automation, i.e., not limited to only RPA.

Forming new roles in the municipality departments. At the different departments, there are digital developers and other stakeholders that are involved in RPA initiatives and participate in the realization of RPA implementations through, e.g., making process descriptions. As a result of the experiences so far, the importance of process ownership at the respective departments has been recognized. The idea is that each automated process should be regarded as a digital colleague and have a process owner—a responsible human colleague. The responsible colleague is knowledgeable about the process (has typically conducted the manual task and contributed to the process description). This model is not yet fully implemented.

The role of external consultants. Initially, three different consultancy firms were involved in the RPA initiatives: one firm supporting the technical infrastructure, and two firms supporting related processes (e.g., RPA initiation, RPA implementation, and competence development of the RPA group). As of 2021, only one consultancy firm supporting related processes is involved. One consultant is involved in the RPA-group and attends weekly meetings.

Forming a process for RPA implementation. Based on experiences from RPA applications and with guidance from the consultancy firm, a work process for RPA implementation was established and revised during 2019–2021. The governance model was not included and remained still to be resolved (during the time of the interviews), but the general idea was that an RPA should have a responsible human colleague in its respective department. Figure 2 depicts the main stakeholder groups involved with different roles and responsibilities relating to RPA within the organization.

As seen in the account above, RPA implementation in Municipality West is situated in a complex web of actors on different levels of the organization affecting the development in different ways. Automation initiatives in the municipality are fueled by politicians and top management. The initiatives are then driven from mid-level roles, many of which have been installed to support RPA initiatives. However, commitment and enthusiasm in the departments is generally lacking. Even though the idea was that RPA should be regarded as a digital colleague rather than an IT initiative, it has been challenging for the RPA group to promote RPA initiatives within the municipality departments. The initial goal was to have 30 RPA applications in use by the end of 2020. In June 2021, three RPA applications were in use and another 5–6 were in development. Most of the initiatives in development were driven by the RPA group with the support of the consultancy firm; proposals were yet to be initiated by the departments themselves. The CIO stated that "*the departments do not really drive digitalization and they are not aware of the potential of automation*" (quote from interview with the CIO for digitalization in 2020). Apart from finding it challenging to inspire the departments to automate processes, they have also run into problems related to staff not wanting to participate in process descriptions. The RPA applications in use have also required more resources than anticipated. Furthermore, they have been accompanied by unexpected delays due to difficulties to access systems, system updates changing the conditions for the applications, and a lack of access to digitized input data necessary for automating processes. This has resulted in ad-hoc development work. The goal for the continued

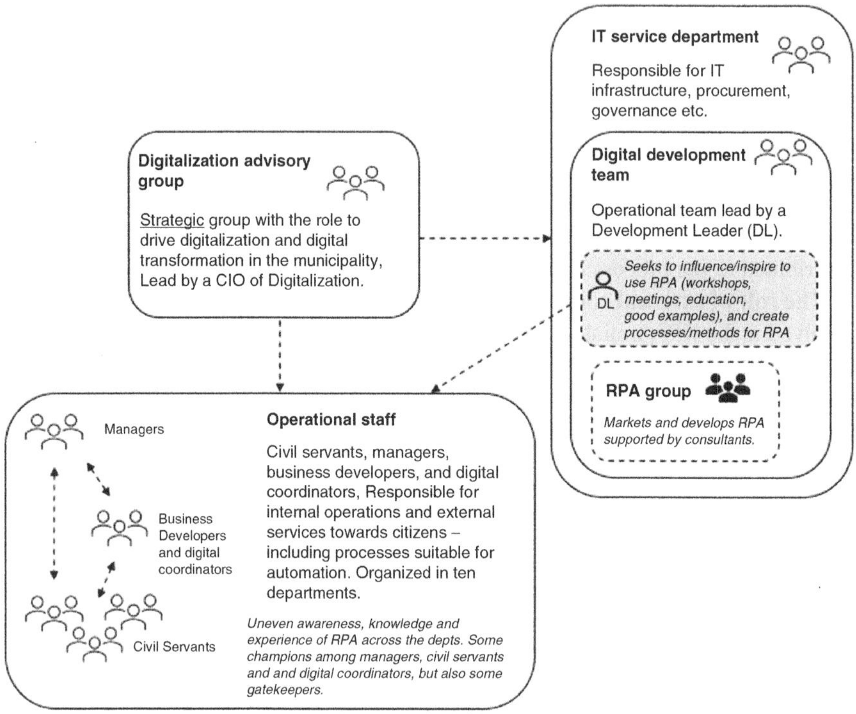

Fig. 2 The main stakeholder groups with their accompanied roles and functions in relation to RPA in Municipality West

development is to finetune the process for RPA implementation and motivate employees to initiate RPA implementation based on needs experienced in the departments.

5 Discussion

The two municipalities' initiatives to organize and implement RPA in their respective organizations share several similarities, but there are also differences. An important similarity between the organizations is that both have chosen to create new roles and structures to prepare for digitalization in their respective organization. However, it appears as if these digitalization structures were not enough to pave the way for RPA implementations; as a result, further restructuring and additional roles were formed to stimulate the specific development of RPA. In municipality West, an operational group within the central IT service department drives RPA initiatives. In contrast, municipality East has deployed a strategic group directly subordinated the Municipal Director to drive RPA initiatives. Thus, in municipality West RPA initiatives is aligned with other digitalization in the organization, while in municipality East there are contrasting perspectives on the merits of RPA dividing the digitalization group and the IT personnel.

Another similarity is that the initiative to implement RPA in the organization comes from actors in the organization that do not work directly in the processes that can be automated. In both organizations, these actors are working actively to inspire those working in the processes to see the potential with RPA and are trying to create a positive "buzz" regarding automation in the organization. Through pilot projects, these actors have tried both to (a) learn about RPA technology and implementation, but also to (b) implement RPA solutions that can function as good examples and inspire co-workers to take initiatives to automate additional processes. With the exception of a few committed enthusiasts in the departments, the take-up of RPA in the departments has been slow in both municipalities. Both organizations have thus created centralized units for digitalization and automation. These structures were created to support employees in the departments who, in turn, were expected to drive the development of RPA based on organizational needs. This arrangement resembles the supportive center of excellence suggested by, e.g., Noppen et al. (2020). However, due to a lack of general interest from the departments, the digitalization units in both municipalities have experienced a need to push and drive RPA initiatives, as also observed by Asatiani et al. (2019). In striving to implement RPA and create good RPA examples that can spur enthusiasm in the departments, both municipalities have conducted their RPA implementations through a more centralized mode of development (see Schmitz et al., 2019) than first intended.

Concerning the actual development and implementation of RPA applications, an interesting difference between the two organizations is that they approach this new territory from somewhat different angles. Municipality East has moved slow with developing actual RPA applications (except for two RPA pilots). Instead, they have focused on creating structures and specified processes for RPA development in the organization, with the aim of scaling up the development of RPA applications once these structures and processes are in place. In contrast, municipality West has started out with practical systems development, moving forward by a trial-and-error approach, creating structures and workflows in parallel. They organized for what could be characterized as an ambidextrous approach (see, e.g., Del Giudice et al., 2021) to prepare for both exploring new RPA initiatives and to make the most of the RPA applications already implemented.

Although the two municipalities have structured their work on RPA slightly differently, both organizations experience the same challenges. First, both municipalities experienced challenges in resource planning. In retrospect, it is possible to see that the frames (time and resources) set for developing, implementing, and then inspiring others to learn about and adopt this technology was overly optimistic in both municipalities. Learning how to work with this kind of technology, getting the pilots going, and being able to implement a functioning RPA application required more resources than anticipated. As a result, the goals set for future RPA development have been altered in both organizations, resulting in less ambitious goals ahead.

Second, both municipalities lacked explicit methods and workflows for RPA development and have been forced to make these up along the way. One area that has been particularly challenging concerns finding processes to automate and creating process descriptions of these processes. In line with previous studies on RPA (see, e.g., Leshob et al., 2018), the case studies illustrate that successful RPA implementation is dependent on creating comprehensive process descriptions. Furthermore, it is important that employees working in the processes are highly involved in defining and modeling the processes (Osmundsen et al., 2019). Both organizations ran into problems concerning the identification of suitable processes to automate as well as engaging employees in process modeling. There are however some examples that contradict this, e.g., in municipality West, where employee involvement in some processes resulted in enthusiasm for the RPA initiative. In municipality East, there is one instance of a business developer who holds an optimistic attitude toward RPA and has started an RPA initiative without being inspired by or formally involved in the initial RPA initiatives.

Last, both municipalities expressed a lack of necessary competences and therefore reported a reliance on external consultants and a few committed enthusiasts in the organizations. Also, the implementation of RPA has triggered discussions on the need for centralization of the IT function, along with discussions on how to create a driving force and demand for digitalization in the departments. The difficulties encountered when implementing RPA applications has also led to a discussion on the role of SALAR and their potential to guide this development.

As a final reflection, models such as the Gartner Hype Cycle[2] seemingly reflect the implementation patterns we observe in both municipalities. Digitalization pressure has generated expectations and interest in RPA. However, the merits of RPA have been exaggerated, resulting in unrealistic expectations of RPA's potential. In both cases, RPA initiatives were legitimized by strategic decisions and initiated by strategic digitalization groups, but then lost momentum and vigor due to (a) slower-than-expected organizational adoption of RPA, (b) absence of visible return-of-investment, and (c) disillusion and impatience by what had been achieved. However, both municipalities are only in the very beginning of their development toward implementing this new generation of automation technologies. If they manage to identify a set of processes where automation truly makes a difference and adds visible value that can be showcased for the employees in the departments, new understanding and optimism is likely to be stimulated.

[2]The Gartner Hype Cycle involves the following steps: (1) Innovation trigger, (2) Peak of inflated expectations, (3) Trough of disillusionment, (4) Slope of enlightenment, and (5) Plateau of productivity. Source: https://www.gartner.com/en/research/methodologies/gartner-hype-cycle

6 Conclusions

We began this chapter by stating that public organizations are in a transformation toward digitalization and digital automation of work. We also noted that purposeful and relevant digitalization is expected to drive quality management and streamlining of public services in Sweden. This trend may have profound implications for public administration, professional discretion, work environment, organizational structure, and services to citizens. Against this background, we have studied how RPA initiatives are organized in Swedish municipalities and reflect on the challenges encountered in this development.

From this study, we conclude that there are significant similarities and differences between how the municipalities have chosen to organize their RPA initiatives. In short, both municipalities have established strategic units to stimulate digitalization and automation on the strategical level. There are however interesting differences in how the municipalities have dealt with the role of their traditional IT departments when organizing new roles and responsibilities to stimulate digitalization, automation, and RPA initiatives. In one municipality, digitalization and traditional IT is kept apart, whereas, in the other municipality, digitalization is subordinated to the traditional IT department. Despite these differences, both municipalities encounter the same types of challenges related to RPA initiatives. Challenges include problems with inspiring employees to take interest in automation; finding suitable processes to automate; gaining access to personnel working in the processes to ensure high-quality process descriptions; a lack of visible return-of-investment; and, differing views amongst stakeholders on the merits of RPA (as a technical solution).

Despite the challenges they have faced, both organizations have managed to implement a set of RPA applications. Interestingly, the successful implementations seem to rely on a combination of using both top-down and bottom-up approaches, where the structures initiated top-down have been met by commitment and enthusiasm from employees in the departments who then took over the responsibility for the RPA implementation and continued the development from a bottom-up perspective. These development projects have been run by a few enthusiasts; however, general interest and commitment is still lacking, making it difficult to move from small- to large-scale implementation.

Our findings suggest that specific factors related to the municipal context shape the organization and development of RPA in the organizations observed. For example, Swedish municipalities are self-governed multi-service providers containing independent authorities, administrations, and heads of administration, as well as professional case handling staff. The distribution of power and responsibilities across and within the organization challenges traditional ways of developing IT and call for alternative organizing principles to realize RPA. In what ways the municipal context shapes the development needs to be investigated further. Also, more research is needed to understand what work practices can help the municipalities to overcome the challenges identified above. From our comparison, we see a need to find new work methods that can resolve (a) the tension between top management's abstract digitalization visions and the experience-grounded ideas originating from e.g., case handling staff, and (b) the tensions springing from responsibilities and power being distributed across RPA stakeholders.

Our research contributes to research on public sector digitalization by providing empirical illustrations of how RPA initiatives can be manifested in local government practice. These empirical illustrations can complement existing research on RPA in the public sector and guide future research on this topic. Our findings also contribute to local government practice. The overviews of each municipality's way of organizing for RPA and the challenges met along the way, together with the comparison between the two municipalities, provide a holistic description of different ways of organizing for RPA implementation. Such a description can provide a useful benchmark for other local government organizations in their efforts to automate work using RPA or similar technologies.

Acknowledgments The research presented in this chapter was financed by AFA Försäkring (AFA Insurance).

References

Åkesson, M., & Thomsen, M. (2020). RPA in service of bureaucracy—When bots are colleagues in everyday public administration. In *Workshop paper presented at the 17th Scandinavian workshop on E-government*. SWEG.

Asatiani, A., Kämäräinen, T., & Penttinen, E. (2019). Unexpected problems associated with the federated IT governance structure in robotic process automation (RPA) deployment. In *Business economy working papers, no. 2/2019*. Aalto University.

Del Giudice, M., Scuotto, V., Ballestra, L. V., & Pironti, M. (2021). Humanoid robot adoption and labour productivity: A perspective on ambidextrous product innovation routines. *The International Journal of Human Resource Management*, 1–27. https://doi.org/10.1080/09585192.2021.1897643

Government Offices of Sweden (2017) För ett hållbart digitaliserat Sverige—En digitaliseringsstrategi. Government report N2017/03643/D.

Hallikainen, P., Bekkhus, R., & Pan, S. L. (2018). How OpusCapita used internal RPA capabilities to offer services to clients. *MIS Quarterly Executive, 17*(1), 41–52.

Klein, H. K., & Myers, M. D. (1999). A set of principles for conducting and evaluating interpretive field studies in information systems. *MIS Quarterly, 23*(1), 67–93.

Lacity, M. C., & Willcocks, L. P. (2016). A new approach to automating services. *MIT Sloan Management Review, 58*(1), 41–49.

Lacity, M., & Willcocks, L. P. (2017). *Robotic process automation and risk mitigation: The definitive guide*. SB Publishing.

Leshob, A., Bourgouin, A., & Renard, L. (2018). Towards a process analysis approach to adopt robotic process automation. In *2018 IEEE 15th international conference on e-business engineering (ICEBE)* (pp. 46–53).

Lindgren, I. (2020). Exploring the use of robotic process automation in local government. In *EGOV-CeDEM-ePart 2020* (p. 249). Published online by CEUR Workshop Proceedings. ceur-ws.org

Lindgren, I., Madsen, C. Ø., Hofmann, S., & Melin, U. (2019). Close encounters of the digital kind: A research agenda for the digitalization of public services. *Government Information Quarterly, 36*(3), 427–436.

Lindgren, I., Toll, D., & Melin, U. (2021). Automation as a driver of digital transformation in local government: Exploring stakeholder views on an automation initiative in a Swedish municipality. In *DG.O 2021: The 22nd annual international conference on digital government research* (pp. 463–472).
Myers, M. D. (2009). *Qualitative research in business and management*. Sage.
Ngwenyama, O., Henriksen, H. Z., & Hardt, D. (2021). Public management challenges in the digital risk society: A critical analysis of the public debate on implementation of the Danish NemID. *European Journal of Information Systems*, 1–19.
Noppen, P., Beerepoot, I., van de Weerd, I., Jonker, M., & Reijers, H. A. (2020). How to keep RPA maintainable? In *International conference on business process management* (pp. 453–470). Springer.
Osmundsen, K., Iden, J., & Bygstad, B. (2019). Organizing robotic process automation: Balancing loose and tight coupling. In *Proceedings of the 52nd Hawaii international conference on system sciences*.
Patton, M. Q. (1980). *Qualitative evaluation methods*. Sage Publications.
SALAR (2018a). Automatisering av arbete—Möjligheter och utmaningar för kommuner, landsting och regioner.
SALAR (2018b). Automatiserad ärendehantering—Att frigöra tid för värdeskapande arbete.
Schmitz, M., Dietze, C., & Czarnecki, C. (2019). Enabling digital transformation through robotic process automation at deutsche telekom. In N. Urbach & M. Röglinger (Eds.), *Digitalization cases*. MP (pp. 15–33). Springer.
Söderström, F., Johansson, B., & Toll, D. (2021). Automation as migration? Identifying factors influencing adoption of RPA in local government. In *ECIS 2021 research-in-progress papers* (Vol. 38).
SOU. (2014). *Automatiserade beslut—Färre regler ger tydligare reglering* (p. 75). E-delegationen.
SOU. (2016). *För digitalisering i tiden* (p. 89). Digitaliseringskommissionen.
Toll, D., Lindgren, I., & Melin, U. (2021). Process automation as enabler of prioritized values in local government – A stakeholder analysis. In H.J. Scholl, et al. (Eds), *Electronic government. Proceedings of the 20th IFIP WG 8.5 international conference, EGOV 2021 (LNCS 12850), Granada, Spain, September 7–9, 2021* (pp. 288–300). Springer.
Trefler A (2018) The big RPA bubble. Article in Forbes, December 2nd, 2018. Available at https://www.forbes.com/sites/cognitiveworld/2018/12/02/the-big-rpa-bubble/.
Walsham, G. (1995). Interpretive case studies in IS research: Nature and method. *European Journal of Information Systems, 4*(2), 74–81.
Willcocks, L. P., & Lacity, M. (2016). *Service automation robots and the future of work*. SB Publishing.
Willcocks, L. P., Lacity, M., & Craig, A. (2015). The IT function and robotic process automation. In *The outsourcing unit working research paper series*. http://eprints.lse.ac.uk/64519/1/OUWRPS_15_05_published.pdf

Managing Two-Speed Innovation. Combining Ambidexterity and Platform-Oriented IT

Bendik Bygstad, Egil Øvrelid, and Robin Williams

1 Introduction

In this article, we propose a framework for governing two-speed innovation. The concept of two-speed innovation builds on two ideas; the research on organisational ambidexterity and the emergence of platform-oriented IT architecture. Organisational ambidexterity means to establish a separate unit to speeden up digital innovation, loosely coupled to the "mother" organisation. This way, an organisation can conduct digital innovation, while maintaining the production of products and services (O'Reilly & Tushman, 2004). The ambidextrous organisation is a two-speed organisation, and we argue that it should be supported by two-speed technologies. During the past decade, IT solutions have developed from monolithic systems to various forms of platform architectures and ecosystems (Tiwana, 2014). Platform-oriented systems are often organised in two layers, a bottom layer of heavyweight systems, such as enterprise solutions and transaction system, and lightweight solutions, such as mobile apps and RPA solutions (Bygstad, 2017). Heavyweight systems evolve slowly, being integrated with other solutions, and subjected to strict requirements of security and privacy, while lightweight technologies are developed quickly, often outside the IT department. The basic idea of two-speed innovation is that ambidexterity and platform-oriented IT are mutually supportive, because they are both directed towards two-speed development.

A key challenge of two-speed innovation is how to manage it, because it requires the interactions of quite different elements. Ambidexterity includes the challenge of

B. Bygstad (✉) · E. Øvrelid
Department of Informatics, University of Oslo, Oslo, Norway
e-mail: bendikby@ifi.uio.no; egilov@ifi.uio.no

R. Williams
School of Social and Political Science, University of Edinburgh, Edinburgh, UK
e-mail: robin.williams@ed.ac.uk

G. Juell-Skielse et al. (eds.), *Service Automation in the Public Sector*, Progress in IS,
https://doi.org/10.1007/978-3-030-92644-1_11

co-ordinating two units, one production-oriented and one innovation-oriented, with very different institutional logics and cultures. Platform-oriented IT must deal with the tensions between a traditional "heavyweight" IT department versus a lightweight innovative unit with different methods, competences and technologies. Combining this in a two-speed configuration is managerially demanding and potentially ridden with conflicts. The problem is illustrated in a Swedish study of the public sector, where the authors concluded that centralised IT governance negatively impacted ambidexterity by systematically skewing the balance between efficiency and innovation (Magnusson et al., 2020). Another issue is how ambidextrous organisations can balance the *installed base* (Hanseth & Lyytinen, 2010) of legacy systems and routines, with disruptive digital innovation. This aspect is not addressed in much depth in current research, but some researchers have identified strong tensions between integration and renewal in IT transformation initiatives (Aakesson et al., 2018; Gregory et al., 2015).

We propose that these challenges cannot be solved by ongoing management, but requires a governance framework. Governance is about "who decides what", i.e. anticipating which issues that will arise, defining the roles to deal with them, and establishing the decision rights (Weill & Ross, 2004). We, therefore, propose a governance framework for two-speed innovation in the next section. Then we illustrate the framework in three cases and finally discuss how managers can use the framework.

2 A Governance Framework for Two-Speed Innovation

Two-speed innovation expresses the twin ideas that an organisation (i) must develop managerial structures and routines to both exploit existing resources and explore new ones (March 1991), and (ii) that platform systems and infrastructures support this by their technical architecture, i.e. having some stable core elements with slow innovation and some user-oriented elements with fast innovation (Baldwin & Woodard, 2009), connected by boundary resources (Ghazawneh & Henfridsson, 2013). Summing-up, the principles of two-speed innovation can be summarised this way.

- It is a process that combines ambidexterity and platform-oriented IT architecture, by a mutually reinforcing configuration: a two-speed organisation, supported by two-speed digital architecture.
- Two-speed innovation allows organisations to leverage their existing resources to enable fast innovation while maintaining their structures and processes.

As shown in Fig. 1, the platform-oriented IT architecture consists of heavyweight and lightweight IT, connected by technical boundary resources (TBL), while the ambidextrous organisation consists of the old and the new business, loosely coupled (O'Reilly & Tushman, 2004). In parallel with the technical boundary resources

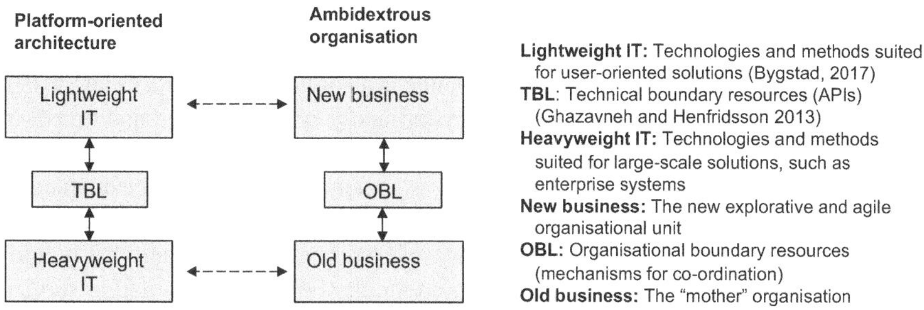

Fig. 1 A framework for two-speed innovation

Table 1 The steps of two-speed innovation

Dynamic capabilities	Platform-oriented IT	
	Heavyweight IT	Lightweight IT
Sensing:	Identifying generative resources	Identifying digital options
Seizing:	Adapting platform and boundary resources	Developing lightweight apps
Reconfiguring:	Mobilising and securing platform performance	Adapting user services

APIs) for the IT architecture we suggest to call this coupling—often the top management group—organisational boundary resources (OBL).

In order to make the framework work for managers, we need to deal in more detail with the interplay of the ambidextrous organisation and the platform-oriented IT architecture. Following O'Reilly III and Tushman (2008) we suggest that the key to operationalise the ambidextrous organisation is to develop dynamic capabilities (Teece, 2012). Dynamic capabilities are needed to leverage digital resources for innovation, particularly to exploit the interplay of heavyweight IT with lightweight innovation (Table 1).

Sensing means, for lightweight IT, to identify digital options, i.e. opportunities to invest in new services that will increase the organisation's value proposition (Rolland et al., 2018). For heavyweight IT, it means to identify digital resources, such as databases, platforms, middleware and security mechanisms, that can be used to support lightweight innovation. This could be transactional data for a possible commercial app, or it could be geographical data for a real-estate service.

Seizing is the step where decisions are taken and investments are made. For lightweight IT this usually includes the development of user services, in the form of mobile apps or IoT services. Heavyweight IT is needed to support these applications, with data and boundary resources. Boundary resources are crucial for enabling API access to the data, and for providing the necessary security and privacy for the users (Ghazawneh & Henfridsson, 2013). For instance, medical records are a valuable data resource for many purposes, but national law and GPRS regulations require a strict governance regime.

Reconfiguring is an ongoing process of realignment and redeployment (Augier & Teece, 2009).

It denotes that digital innovation and change is a continuous process (Vial, 2019), it implies that a platform ecosystem must continually be orchestrated and tuned, in order to work effectively (Eaton et al., 2015; Ivarsson & Svahn, 2020). Lightweight apps must interact with other services, such as social media and must be adapted to changing user needs. The role of heavyweight IT is quite different; it must ensure that the platform services maintain integrity and operate smoothly, and that boundary resources are adapted to changing requirements. For instance, a dating service must monitor the rate of successful matchings, and adjust both the app logic and the database (Parker et al., 2016).

3 Method

To validate and illustrate the framework of two-speed innovation, we adopted a comparative case study method and chose three *typical* cases, which are considered to be suitable for theory assessment (Gerring, 2007). Selection criteria were (i) that the case included both an ambidextrous organisation structure and a platform-oriented digital solution, and (ii) with a reasonably successful outcome. Further (iii) we chose cases that included longitudinal data, in order to assess the development of dynamic capabilities over time. We carefully selected three example cases, from our portfolio of ongoing longitudinal studies of innovation in large interconnected structures:

- Østfold Hospital: A large general hospital
- TSD—by USIT: A research platform at the University of Oslo
- Nordic Choice: A Scandinavian hotel chain

Data was collected over the period 2015–2021, consisting of interviews with key stakeholders, archival data, and observations (see Table 2). The interviews were semi-structured, focusing on digitalisation processes, seen from the perspective of the developers and the user. In the Østfold case, the users were mainly clinicians; in the TSD case, the users were researchers, while in the Nordic Choice case, the users were analysts and hotel staff.

Table 2 Data collection

Case and period	Interviews	Documents
Østfold Hospital 2015–2019	More than 40 interviews: clinicians, managers, IT personnel	Project plans and reports Architecture documents
TSD—at University of Oslo, 2019–2021	11 interviews: managers, IT personnel, users	Project plans and reports Architecture documents
Nordic Choice 2017–2020	25 interviews: managers, IT staff, hotel staff	Project plans and reports, booking statistics Architecture documents

Fig. 2 Two-speed innovation at Østfold Hospital

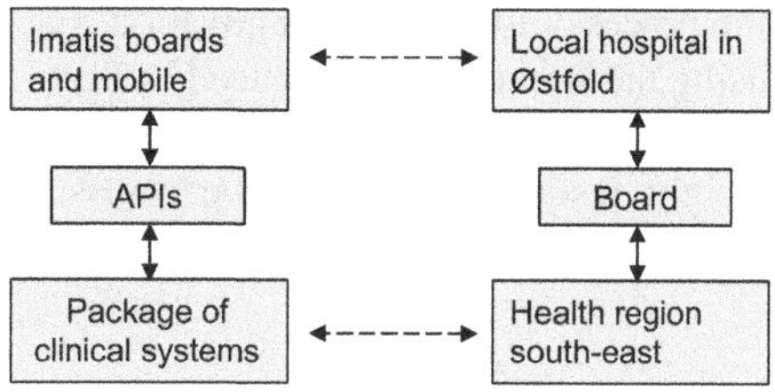

Data analysis was initially conducted for each case; using our framework we analysed the interplay of heavyweight and lightweight IT through the three phases of dynamic capabilities, i.e. sensing, seizing, and transforming. In particular, we assessed the different roles of heavyweight and lightweight IT and the outcome of the innovation process. Then we conducted a comparative analysis of the three cases, aiming to identify similarities and differences.

Following Gregory et al. (2015) we also investigated the tensions in the cases, focusing on both organisational and technical relationships. For instance, in the Østfold Hospital case, there were strong initial tensions between the local team and the IT Centre, where the IT Centre questioned both the need for lightweight solutions and the integration shortcuts that the project proposed. In parallel, the Østfold Hospital CEO met resistance in the regional authority for "branching out" of the official strategy.

From these analyses, we derived a more general model, involving particular configurations of tight and loose couplings (Fig. 2), which helps to explain the forces in two-speed innovation. We used this systematically to re-assess the cases and identify the managerial challenges of the configuration.

4 Three Cases

The two-speed configuration is an ideal type, in a Weberian sense, and one should not expect to find it in pure form. Rather, our interest is to explore the forces and challenges within it. To do so, we use the three cases, shown in Table 3, to describe

Table 3 Cases

Case	Description	Ambidextrous organisation	Platform-oriented IT architecture
Østfold Hospital	General hospital	Local hospital initiative vs. regional health authority	Lightweight patient logistics solution vs large clinical systems
University of Oslo	Research platform for University of Oslo	TSD group vs. USIT (IT dept.)	Agile app architecture vs. central storage
Nordic Choice	Large Scandinavian Hotel chain	Innovative unit eBerry vs. large hotel org.	Apps and analytics vs. booking systems

the process of two-speed innovation. Each case description is structured as follows, using the framework introduced in Sect. 2; we analyse the steps of sensing, seizing, and reconfiguring, focusing on the interplay of lightweight innovation and heavyweight resources. Finally, we assess the outcome of the two-speed innovation process.

4.1 Østfold Hospital

In 1999, the Norwegian Parliament decided to build a new hospital in Østfold County, which is part of the South-East Regional Health Authority. The hospital opened in November 2015, with both somatic and psychiatric services and 4800 employees. It quickly became well-known as one of the most digitally advanced hospitals in Scandinavia.

4.1.1 Sensing

Hospitals have historically been organised according to the division of medical specialisations and have been less occupied with the coordination of activities across these disciplines. This has resulted in many heavyweight IT silo solutions, and the South-East Health Region had standardised a portfolio of around 300 such systems. These systems were traditional "silo systems", i.e. electronic patient record, lab, radiology, and other clinical solutions, designed to support specific user groups. They supported well these purposes but provided little support for the flow of patients through the various units of the hospital.

However, the CEO at Østfold hospital, Just Ebbesen, had other ideas. He was a doctor and a pioneer in using lightweight IT to innovate and support clinical processes: "I had been engaged with the relationship of process innovation and IT the past 15 years, both theoretically and practically, and I knew what I wanted to achieve: hospital processes should be well defined and supported by information."

4.1.2 Seizing

In 2014 Ebbesen hired a CIO with experience from production and retail and a process director and established a new top management team. The team decided to acquire process technology and contacted a new supplier, Imatis. This process technology was not part of the regional portfolio of heavyweight clinical systems. An innovation project was established, with around 25 clinicians working on redesigning the clinical processes, and a separate group that worked with the process technology. The IT systems had to relate to the process design, and since much of the information was stored in the existing digital infrastructure (consisting of over 300 applications, and managed by the central IT unit), an interface that the process technology could use was created. The process technology consisted of lightweight applications, check-in automats, mobile devices, and electronic whiteboards, i.e. user services that enabled clinical personnel to configure the information to fit with the process flow.

Implementation was quite demanding. When the lightweight vendor contacted one of the heavyweight vendors for access to data the answer was, "well, our system does not offer an API, and there is nothing in our contract that allows for 3rd party access to our data". It took great efforts, and escalation to top managers in three organisations, to solve this problem. In addition, the cooperation with the regional IT Centre, was full of tensions; the IT staff at HospitalPartner placed the requests from the Østfold IT project in the regular queue, which was long. The staff also did not understand the aims of the project and regarded the integration requirements as an inconvenience. These issues were eventually resolved with improvisation and managerial interventions, but the IT Centre staff resented the technical shortcuts of the lightweight project.

4.1.3 Reconfiguring

The endeavours to implement lightweight process IT, and establish interaction with the digital infrastructure, has made Østfold a two-speed organisation. Further innovation included new lightweight initiatives. A new analytics unit was established in 2018 to monitor the performance, and new digital services were created. In 2019–2020 the hospital engaged in an innovation project which enabled cancer patients to receive treatment at home, building on a lightweight solution from another company, Dignio.

4.1.4 Outcome

The two-speed configuration of Østfold Hospital, with tight couplings between the local management and the lightweight solution (and the lightweight vendor), was successful and earned the hospital wide recognition.

However, the relationships to the health region (Health South-East) and the regional IT department proved more challenging. The lightweight solution had been financed mainly through the building budget, and the regional authority was not willing to grant more resources to Østfold than to the other hospitals in the region. This made further developments in Østfold dependent on other sources, such as the Research Council. In the same line, the regional IT Centre remained sceptical of the technical solutions at Østfold. This illustrates that the two-speed configuration was tolerated, but not really accepted at the regional level, meaning that the configuration was much harder to stabilise in the looser couplings.

4.2 *University of Oslo*

The second case is taking place at the University of Oslo, which is the highest-rated university in Norway, with 28,000 students and 6000 employees. The unit of analysis is USIT, the University of Oslo's central IT organisation, with around 220 employees, and thus one of the largest state in-house IT organisations in Norway. USIT delivers a range of IT services, including student record and exam systems, office systems, and high-performance computing, both locally, nationally, and internationally. Half of the earnings are from external customers.

4.2.1 Sensing

Around 2012, USIT experienced an increasing demand for a place to store large volumes of sensitive research data, particularly from clinical and hospital researchers working with quantitative investigations. Further, there was an increasing demand from researchers to create apps on top of the research data. This was a new area for USIT, and a group was established to work on the development of a research platform. A general requirement was that TSD should as far as possible use the hardware and software resources that already existed, and that the operation should take place as part of USIT's common operating environment.

However, USIT's competence structure indicated that they lacked innovation expertise. A competent project manager was found and appointed, and a small group was established. Stronger ties were also established with important research groups at the University.

4.2.2 Seizing

Initially, the TSD group had to use the existing integration engine when developing new solutions. However, it was not designed in a way that made it easy to create new apps on top, being a tightly coupled to underlying systems. The TSD group, therefore, implemented a new architecture to facilitate the establishment of a more flexible platform-oriented solution. Central to this initiative was a new architecture developed for providing flexible and secure transport and data management services The new architecture, provided (i) boundary resources, i.e. API semantics for offering multi-tenant services, (ii) a secure and modular architecture, (iii) a flexible model for implementing access control, and (iv) a development framework for the implementation of application servers and API clients. TSD thus developed towards a platform architecture consisting of a platform core, boundary resources, and loosely coupled app modules (see Fig. 3).

TSD was launched as a private cloud service, which quickly gained popularity, used by more than 1000 research projects. It offered a variety of application resources that allow services to be developed by the researchers themselves.

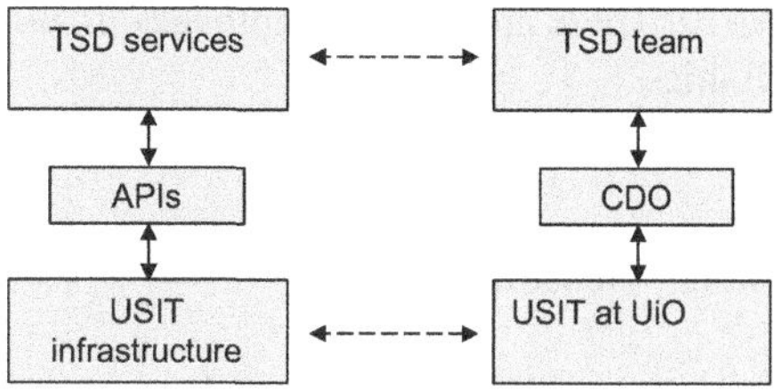

Fig. 3 Two-speed innovation at University of Oslo

Typically, the projects used DNA sequencing, web-based questionnaires, smartphone applications, and therapy sessions video streaming. For instance, the TSD group responded early to the corona crisis in March 2020 and quickly provided researchers on the pandemic with a secure service (ref 2020).

4.2.3 Reconfiguring

TSD eventually became central to USIT's innovation strategy. Realising the potential of the solution, TSD was in 2019–2020 extended to a general research platform for all researchers at the University of Oslo, and possibly a national solution. The solution built on the FAIR principles (findable, accessible, interoperable, reusable), and was competing with other international solutions in the higher education and healthcare sector.

4.2.4 Outcome

With TSD, USIT established a two-speed configuration, enabling the organisation to manage both stable operations and resilient innovation. By developing and implementing TSD, USIT created a platform-oriented architecture, with the database solution tightly coupled to the USIT heavyweight systems. Boundary resources provided secure and flexible access to a range of digital resources within various heavyweight systems. USIT was also through TSD been able to establish a robust "daughter" organisation for facilitating digital innovations. Although this occasionally led to cultural challenges in a traditional organisation, the change was implemented without significant conflicts. One reason was that the research communities paid for the development of services.

4.3 Nordic Choice

Nordic Choice is a hotel chain based in Scandinavia. It includes 205 hotels and 33,500 rooms and had a turnover of 12 billion NOK (around 1.3 billion euros) in

2019. There are 17,000 employees in Norway, Sweden, Denmark, Finland, and the Baltics.

4.3.1 Sensing

In the late 1990s Internet-based online travel agencies (OTAs), such as Booking.com and Hotels.com became popular as convenient intermediaries for hotel customers. For years, Nordic Choice had regarded the OTAs as valuable partners that offer an effective marketing and distribution channel. In 2014, however, the management team realised that the increasing number of bookings through OTAs led to price pressures and lower margins. Moreover, by dealing directly with the customers, the OTAs accumulated the information Nordic Choice needed to manage customer relationships. As expressed by the company's vice president:

> First, the online booking websites, such as Hotels.com and Booking.com, emerged. We then regarded them as helpful add-ons, making it easier to find us, but got worried when their share of the room price got greedy. Then other services emerged, such as TripAdvisor, placing themselves between the customer and the booking websites, also making money on our customers. Over time, an increasing share of our customers communicated with these sites, and not with us. I realized that if nothing were done, we would end up as a commodity provider of hotel rooms, leaving the distribution to the internet companies.

Responding to the threat, Nordic Choice initiated a digital business strategy. The hotel chain realised it lacked the resources to implement the strategy, and established a separate company, eBerry in 2016. The mandate of eBerry was to maintain the main share of the bookings of Nordic Choice in the distribution chain.

4.3.2 Seizing

A Chief Digital Officer was hired with a background from one of the OTAs. She quickly engaged 40 new IT experts and established several teams to compete in the digital ecosystem. eBerry started to implement the strategy short-term by entering the digital competition arena (negotiating with OTAs, search optimisation, continuous monitoring of digital traffic, and long-term by investing in new digital solutions. These included a technical platform to build new solutions on, (see Fig. 4), but integrated with the legacy booking systems, a new website, and a loyalty app.

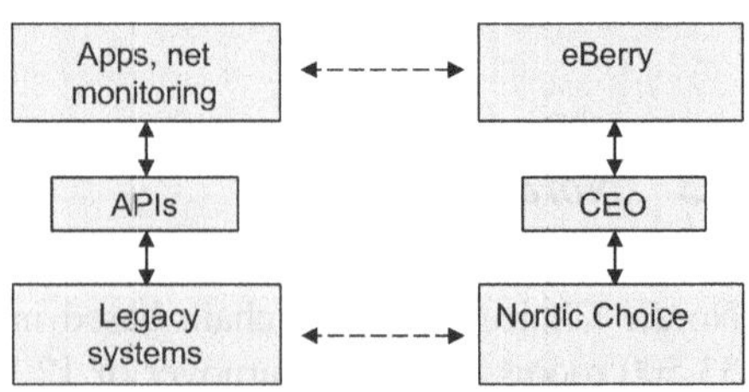

Fig. 4 Two-speed innovation at Nordic Choice Hotels

The culture of eBerry was very different from the mother company. Said the CIO: "We do not run projects; we aim to be a technology firm building products. We have three main products: the Nordic Choice website, the integration platform, and the Nordic Choice app. We run the distribution for Nordic Choice, with a turnover of 5 billion NOK each year. We are responsible for the loyalty program, campaigns, and bookings. We also have a 'Future Business' unit developing new technologies and ideas, such as robots."

The results were good; these measures stabilised the Choice share of bookings, and they reduced distribution costs significantly. A key element was the *second-visit* strategy. Choice accepted that many customers would book with the OTAs. However, when the guest arrived at the hotel, the receptionist would always offer membership of the bonus programme, including an upgrade of hotel services, and an app. The app offered convenient booking and check-in services and was later also designed to be used as a room key. The aim was to ensure that the customer would book directly for the next visit.

4.3.3 Reconfiguring

The OTAs frequently change their business tactics as competitors are bought, new services are launched and new actors arrive. At the strategic level, the Nordic Choice top management group and the board frequently assessed the OTAs' activities and services, allowing them to continuously consider and reconsider their strategic positioning in the digital hospitality ecosystem. At the tactical level, managers focused on the communication channels directly with customers using e-mail marketing, the website, and the app. A broad campaign in 2018 encouraged travellers to book with Nordic Choice directly. On the operational level, one eBerry team specialised in surveying the web and booking traffic 24/7 to follow the competition arena in real-time. This information was analysed daily and used systematically to improve performance.

4.3.4 Outcome

The strategy established a fast-moving unit, eBerry, with specialised competence and a different culture. It innovated quickly a set of digital services, which were integrated relatively loosely with the legacy heavyweight solutions at the mother company. In managerial terms, there was also loose coupling, as the cooperation between the two units was mainly done at the top management level.

5 Discussion: How Can Managers Leverage Two-Speed Innovation?

Our three cases show that two-speed innovation, the combination of ambidexterity and platform-oriented IT, can be governed by a framework that mitigates the contradictory forces, and allows for fast digitalisation. Digitalisation is a challenging task for managers of most incumbent organisations (Gregory et al., 2015). As observed by Dumaine, "the worst way to speed up a company['s] [innovation process] is by trying to make it do things just as it does, only faster. The machinery, and certainly the workers, will simply burn out" (Dumaine, 1989), p. 55.

Our cases highlight that two-speed innovation manages this tension, but it requires a careful strategic assessment from the top management, and skilful implementation from the middle managers. Often, organisations simply lack the necessary managerial skills to conduct the analysis and develop dynamic capabilities. For instance, in the case of Nordic Choice, the top managers did not have the needed insights into the OTA ecosystems and decided to recruit the CEO from eBookers, a platform company. She managed the establishment of eBerry and also became a member of the top management group of Nordic Choice. Similarly, the innovation at Østfold Hospital was conducted as a joint initiative between a new visionary CEO and a new CIO with logistics experience.

Our cases illustrate some of the preconditions for succeeding with two-speed innovation. Two-speed innovation requires that top managers, in the *sensing* phase, understand and leverage the idea. Developing the necessary dynamic capabilities implies risk developing resources that the organisation currently lacks (Teece, 2012). Taking the example of Østfold Hospital the CEO initiated a process view of the new hospital, and the local team implemented a completely new solution. In the case of USIT, the CIO decided to allocate key resources from operations, and start building a research platform.

In the *seizing* phase of two-speed innovation, the role of middle managers is crucial. It is essential to leverage the interplay between heavyweight and lightweight IT, to support the ambidextrous organisation (O'Reilly III & Tushman, 2008). The reason is that tight integration between the two will slow innovation speed, reducing it to the speed of heavyweight IT (Bygstad, 2017). In the case of Nordic Choice, the new unit, eBerry, established agile processes and culture very different from the hotel chain, enabling fast development of new digital resources.

Finally, our examples show how the *reconfiguration* phase is critical for success. Digital innovation and change is not an end-state but a continuous process (Aakesson et al., 2018). As the Nordic Choice case illustrates, a disruptive lightweight solution is not a finished product, but the first version of a long series of technical improvements and new customer services.

6 Conclusion

This paper introduced and discussed a governance framework for two-speed innovation. We show that digital innovation and change is made possible by a combination of ambidextrous organisation and platform-oriented IT architecture. We explored how the contradictory forces of two-speed innovation played out in three cases in large organisations. Finally, we discuss how managers can leverage two-speed innovation for organisational purposes, and find that managers should deal with the challenges of two-speed innovation by carefully developing and executing the needed dynamic capabilities and by developing actors able to bridge between and deal competently with frictions between the units.

References

Aakesson, M., Sørensen, C., & Eriksson, C. I. (2018). Ambidexterity under digitalization: A tale of two decades of new media at a Swedish newspaper. *Scandinavian Journal of Management, 34*(3), 276–288.

Augier, M., & Teece, D. J. (2009). Dynamic capabilities and the role of managers in business strategy and economic performance. *Organization Science, 20*(2), 410–421.

Baldwin, C. Y., & Woodard, C. J. (2009). The architecture of platforms: A unified view. *Platforms, Markets and Innovation, 32*.

Bygstad, B. (2017). Generative innovation: A comparison of lightweight and heavyweight IT. *Journal of Information Technology, 32*(2), 180–193.

Dumaine, B. (1989). How managers can succeed through speed. *Fortune, 119*(4), 54.

Eaton, B., Elaluf-Calderwood, S., Sørensen, C., & Yoo, Y. (2015). Distributed tuning of boundary resources: The case of Apple's IOS service system. *MIS Quarterly, 39*(1), 2017–2243.

Gerring, J. (2007). *Case study research: Principles and practices*. Cambridge University Press.

Ghazawneh, A., & Henfridsson, O. (2013). Balancing platform control and external contribution in third party development: The boundary resources model. *Information Systems Journal, 23*(2), 173–192.

Gregory, R. W., Keil, M., Muntermann, J., & Mähring, M. (2015). Paradoxes and the nature of ambidexterity in IT transformation programs. *Information Systems Research, 26*(1), 57–80. https://doi.org/10.1287/isre.2014.0554

Hanseth, O., & Lyytinen, K. (2010). Design theory for dynamic complexity in information infrastructures: The case of building internet. *Journal of Information Technology, 25*(1), 1–19.

Ivarsson, F., & Svahn, F. (2020). Becoming a digital ecosystem orchestrator-The Sydved Case.

Magnusson, J., Khisro, J., & Melin, U. (2020). A pathology of public sector IT governance: How IT governance configuration counteracts ambidexterity. In *International Conference on Electronic Government* (pp. 29–41). Springer.

March, J. G. (1991). Exploration and exploitation in organizational learning. *Organization Science, 2*(1), 71–87.

O'Reilly, C. A., & Tushman, M. L. (2004). The ambidextrous organization. *Harvard Business Review, 82*(4), 74–81.

O'Reilly III, C. A., & Tushman, M. L. (2008). Ambidexterity as a dynamic capability: Resolving the Innovator's dilemma. *Research in Organizational Behavior, 28*, 185–206.

Parker, G. G., Van Alstyne, M. W., & Choudary, S. P. (2016). *Platform revolution: How networked markets are transforming the economy? And how to make them work for you*. WW Norton & Company.

Rolland, K. H., Mathiassen, L., & Rai, A. (2018). Managing digital platforms in user organizations: The interactions between digital options and digital debt. *Information Systems Research, 29*(2), 419–443.
Teece, D. J. (2012). Dynamic capabilities: Routines versus entrepreneurial action. *Journal of Management Studies, 49*(8), 1395–1401.
Tiwana, A. (2014). *Platform ecosystems*. Morgan Kaufmann.
Vial, G. (2019). Understanding digital transformation: A review and a research agenda. *The Journal of Strategic Information Systems*.
Weill, P., & Ross, J. W. (2004). *IT governance: How top performers manage IT decision rights for superior results*. Harvard Business Press.

What Can Public Sector Organizations Learn from Private Sector Experiences of Robotic Process Automation?

Aleksandre Asatiani

1 Introduction

Robotic process automation (RPA) has experienced phenomenal growth during the last decade, rising from a potentially promising new technology in 2012 (Fersht, 2012) to a US$1.58 billion industry in 2020 (Costello & Rimol, 2020). RPA was hailed as a fast-to-deploy and cost-effective automation tool that can be applied to a wide range of routine tasks. Early adopters reported hundreds of full-time equivalents (FTE) saved thanks to RPA (Lacity & Willcocks, 2016), but as more organizations have adopted RPA, the promised benefits have become less certain (Willcocks, 2019). It has become clear that the promised benefits of RPA are contingent upon its successful deployment, and while it is quick and simple to deploy, the success of an RPA initiative is not just a function of using the technology correctly; as with other enterprise technologies, RPA requires planning and active management.

In the 2019 edition of Gartner's Hype Cycle, the company placed RPA in its *trough of disillusionment,* indicating that the hype for RPA was over (Sicular et al., 2019). Many users and would-be users of RPA started to adjust their expectations as the best practices for RPA deployment and management started to emerge; some even started to ask whether RPA is still relevant. That being said, the end of the hype appears to be a sign that RPA is reaching maturity rather than the end of its lifecycle. There is a strong track record of successful RPA deployments (Lacity & Willcocks, 2021), and it is also becoming part of the standard IT toolkit in organizations, serving as a foundation for more advanced *intelligent automation* (Willcocks, 2021).

The majority of RPA early adopters were private sector organizations (Kroll et al., 2016), and it only started to appear in the public sector relatively late but

A. Asatiani (✉)
Swedish Center for Digital Innovation, University of Gothenburg, Gothenburg, Sweden
e-mail: aleksandre.asatiani@ait.gu.se

G. Juell-Skielse et al. (eds.), *Service Automation in the Public Sector*, Progress in IS,
https://doi.org/10.1007/978-3-030-92644-1_12

spread quickly (Ranerup & Henriksen, 2020). For the public sector, RPA is seen as an effective way to automate manual, rule-based tasks and as a foundation for a gradual approach to introducing more complex automation tools, such as artificial intelligence (AI) (Berryhill et al., 2019). Arguably, the RPA hype is still ongoing in the public sector, particularly in the Nordic region. However, being a latecomer gives the public sector the benefit of hindsight, as it can learn from the experiences of private sector organizations and avoid some of the same mistakes. In this chapter, I will highlight some of the contemporary challenges and lessons learned from deploying RPA in the private sector that have been identified in contemporary RPA research. My own work on RPA focuses on private sector organizations, and my aim here is therefore to focus more on the lessons learned from the private sector rather than to produce prescriptions for public sector organizations.

2 RPA Initiatives Are Easier to Start Than to Scale

One of the key advantages of RPA is how easy it is to learn, develop, and deploy (Lacity & Willcocks, 2016). Typically, it takes anywhere from two to six weeks from inception to deployment (Asatiani & Penttinen, 2016; Syed et al., 2020), which is significantly faster than the deployment of most other automation tools. Such easy deployment also means lower costs than conventional IT projects, which makes RPA an attractive option because the barriers to entry are very low. Organizations often go into RPA with a lot of optimism and a long list of routine tasks to be automated, but even after a successful proof-of-concept (PoC), RPA initiatives can still stall and fail to scale, despite seemingly positive results (Lacity & Willcocks, 2021; Willcocks, 2021). Some in the industry have started to refer to these as *PoCs of death*, and it appears that deploying RPA to automate a specific process is quite different from deploying tens or hundreds of robots across the organization. The problem of scaling is one of the major contemporary challenges of RPA, and both research and industry best practices provide insight into these challenges and offer useful lessons.

Selection of the first PoC is extremely important, and the prevailing wisdom is to think big and start small (Syed et al., 2020). Reaching for the lowest hanging fruit for a PoC is logical for several reasons. It is in the interests of the organization to start the RPA initiative on a high note, and such an approach allows for quick and easy wins, reduces the risk of a failed PoC, and helps to avoid a *scope creep* trap. At the same time, not all low-hanging fruits are the right choice for a PoC; a process selected for automation needs to have an observable positive impact and to deliver tangible value to the target group of stakeholders within the organization. Automating processes that hardly anyone encounters in their day-to-day work could create the perception that RPA has only limited capabilities (Lehtinen et al., 2020), even if—on paper—the PoC has achieved significant efficiency gains. The main purpose of a PoC is to create a strong business case and to "sell" the idea of RPA to the rest of the organization. The PoC, therefore, needs to be enticing in a broad sense, beyond dry figures for efficiency gains and cost savings.

Scaling RPA requires stakeholder buy-in across the organization; for a pilot project, it is relatively easy to get such buy-in from both higher management and workers of a single unit—there are often enthusiastic groups within IT departments and back-office units (e.g., accounting) who are willing to experiment with new automation technology to eliminate unpopular routine tasks done by human workers. However, scaling RPA beyond these early PoCs requires much broader support. Managers of other, more diverse units may not be persuaded by an RPA business case (Lacity & Willcocks, 2021) and may refuse to support an RPA initiative in their units. Operational workers, on the other hand, may feel threatened by the potential for automation to take their jobs (Asatiani et al., 2020b), leading to resistance and backlash. In cases in which an IT department is not involved at the early stages of RPA, it might refuse to cooperate with scaling of the initiative. Buy-in from these stakeholders is essential for the potential of RPA to be utilized throughout the organization, and a lot of preparatory groundwork, therefore, needs to be done in addition to a successful PoC to ensure that the RPA initiative gains broader traction.

RPA also requires a champion inside of the organization in order to scale successfully. Developing and deploying a PoC may not require a dedicated person to push the project forward, given the short duration and straightforwardness of such projects, and external consultants are usually engaged in their development and deployment and, together with project initiators, ensure the project is successfully completed. However, scaling the RPA initiative cannot be someone's low-priority side-job. The RPA champion will spend considerable time resourcing and supporting the project and needs to have a thorough understanding of the *business* side of the organization, high levels of credibility, and a proven track record of delivering projects in business and/or IT (Willcocks et al., 2019). The job of RPA champion includes securing sufficient resources for the RPA initiative, ensuring there is a steady pipeline of RPA projects, and evangelizing RPA within the organization.

3 RPA Requires Long-Term Thinking

RPA may come with a promise of quick wins, but research suggests that successful RPA initiatives tend to be treated as strategic (Lacity & Willcocks, 2021). Organizations considering RPA typically have many processes that can be automated across different units and functions, but decision-making regarding RPA deployment should be driven by the business strategy (Syed et al., 2020; Lacity & Willcocks, 2021). An organization deploying RPA should have a vision of what it wants to achieve with the technology, where RPA can deliver the most value, and how this fits with the overall objectives of the organization. The strategic vision should then inform more specific decisions regarding RPA.

Organizations need to ensure alignment between their strategy and the RPA operating model. When deploying RPA, managers need to decide how it will be operated, which can be distilled into three major questions: (1) Who will develop and maintain the RPA? (2) Will RPA be deployed on the premises or in the cloud? (3) What type of RPA software will be used? Using external RPA developers, cloud-based deployment and off-the-shelf RPA software could enable rapid scaling because most of the resources will come from outside the organization. However, that would also take away the organization's control of the technology and its development, which could prove important in the long term. Conversely, developing and deploying RPA internally may allow for greater control and the ability to closely integrate RPA with an overall IT landscape in the organization, but the shortage of RPA talent and the extra resources required to deploy RPA fully in-house could prove to be a major obstacle to the success of an RPA initiative. Organizations deploying RPA need to carefully evaluate their operating model and keep updating it as the RPA initiative grows and strategic priorities change.

Organizations also need to find the right balance between internal and external resources. RPA is still a relatively new and fast-growing technology, and the majority of organizations starting their RPA journey lack relevant internal competences. Industrial reports suggest that there is a shortage of RPA talent (Peña, 2021), making hiring skilled RPA professionals hard. This means that, at least initially, external RPA consultants and developers will be essential. At this stage, external RPA consultants are more likely to deliver a faster, more cost-efficient, and lower-risk solution to get things off the ground, but in the long term, organizations need to evaluate the internal and external resources that they require to realize their strategic vision. In discussing scalability, we already touched on the need for an internal RPA champion. In addition to such a champion, the organization may consider building internal RPA competences, which could be acquired gradually, either through training existing staff (e.g., software developers, project managers) or by hiring RPA professionals from outside. Having internal resources may deliver more stability as external consultants may change more frequently. Internal RPA developers and managers will likely also have a better grasp of organizational processes and strategic priorities, and in-house RPA staff will be quicker to respond to urgent RPA needs that may arise, such as disruptions introduced by changes to the process or to the systems with which the robots interact.

An organization's strategic RPA vision should also consider the changes that the technology brings to its workforce. In the early days of RPA, there were concerns about it causing large-scale job losses (Lacity & Willcocks, 2015; Asatiani et al., 2020b), but these do not appear to have materialized, due partly to the fact that RPA generally automates only the routine parts of a job, rather than fully replacing a human. Consequently, organizations have generally either reassigned workers whose routine tasks were automated to more value-added work or gradually reduced the workforce through attrition, rather than active downsizing. This does not mean that RPA has had no significant impact on the workforce, because the same positions after the introduction of automation will likely require a different set of skills than before (Asatiani et al., 2019; Lacity & Willcocks, 2021), and the humans that rely on software robots to perform routine parts of a job would need to focus on the skills required for the non-routine tasks. Moreover, such people would likely need to have skills to manage and perhaps program the robots. Organizations deploying automation at scale, therefore, need to revise their job descriptions and prepare their HR departments for the coming of hybrid human–machine jobs.

4 RPA Is Not All About FTEs

Decisions about the suitability of RPA for a specific task often come down to the potential FTE savings. RPA developers and consultants typically look for processes that contain enough routine tasks suitable for RPA automation to justify the project. Processes that save too few FTEs are considered not worth pursuing, but conversely, processes that are too FTE-intensive may be better suited to more heavyweight automation (Bygstad, 2016). However, both academia and industry have recently and gradually started to shift from a discussion about FTEs to one about *value delivered* (Lehtinen et al., 2020) and *hours back to the business* (Lacity & Willcocks, 2021).

There are two reasons for this. First, FTE-dominated thinking draws attention away from the strategic dimensions of RPA and toward rapid efficiency gains and cost-cutting. As discussed above, ignoring strategy can damage an RPA initiative in the long term and ultimately minimize the potential gains from automation. Second, a focus on FTEs emphasizes cuts in the human workforce rather than an increase in the value of their work. Software robots can only perform mindless, routine tasks that represent only a part of a person's job, and there is, therefore, a lot more to be gained from refocusing workers on their core, value-generating responsibilities than from eliminating their jobs completely. Such thinking can be especially relevant to public sector organizations that do not have to cut costs for the sake of improved profitability. For example, in the public healthcare context, saving mere minutes each day on routine administrative tasks for nurses and doctors could, while unimpressive on the surface, mean an extra patient receiving care that they otherwise would not.

Nevertheless, the financial viability of RPA projects remains very important. Overall, software robots are cheaper than human labor, but the deployment and maintenance of RPA does come at a cost, so automating projects with low potential FTE savings will likely be counterproductive. At the same time, organizations evaluating processes for RPA suitability should take a broader view of the value of automation, going beyond FTEs saved and costs cut.

5 RPA Comes with Accountability Concerns

As automation becomes a more integral part of organizational processes in both the public and private sectors, there are growing worries about accountability (Martin, 2019; Asatiani et al., 2021). Many decisions are becoming fully or partially automated, and there are concerns regarding the responsibility for errors and damages caused by the actions of automated agents (Rinta-Kahila et al., 2021). These concerns are being reinforced by regulatory initiatives, such as the European Union's General Data Protection Regulation (GDPR), which grants citizens the right to an explanation of any decision based on data gathered about them (European Union, 2016). While the accountability discussion often focuses on AI and machine learning, which notoriously suffer from explainability problems (Rosenfeld & Richardson, 2019; Asatiani et al., 2021), rule-based automation tools like RPA are not immune from similar dangers. For example, Finland's Parliamentary Ombudsman recently declared the Finnish Tax Administration's rule-based automated system for handling taxation decisions unlawful, determining it to be inconsistent with the principles of good governance, due process, and accountability (Lindström & Sakslin, 2019).

Public sector organizations considering RPA need to take the utmost care to ensure that automating a specific process does not introduce accountability hazards or negatively impact society. First, the organization should determine whether a process has a direct impact on the stakeholders and evaluate the severity of any negative effects caused by errors in the process. For example, if RPA were to ultimately be providing input into life-or-death decisions or handling highly sensitive personal data, one may reconsider deploying it, regardless of a solid business case for automation. Second, the organization should consider the design approaches used in AI and privacy-focused information systems. For example, the concept of *privacy by design* (Cavoukian, 2009) describes information systems that put data privacy first, building the rest of the system around this foundation. Another emerging approach from AI research that may suit RPA is envelopment (Robbins, 2019; Asatiani et al., 2020a), which advocates for a system design that sets clear boundaries for automation to ensure human control over the important decisions and operations of the AI. In the context of RPA, this may mean designing automation with multiple software robots that sandbox private data from the rest of the process, passing forward only necessary pieces, instead of giving access to all the data to a single robot. Such a design would also engage humans in making important decisions and provide the necessary information to do so.

6 Conclusion

After explosive growth, RPA is slowly transforming from a hyped emerging technology to a mature tool for automation. In my recent conversations with practitioners, I have observed the narratives surrounding RPA gradually becoming about a tool in a standard IT toolkit that can be deployed as needed alongside other lightweight and heavyweight automation. At the same time, challenges remain, and a lot of early assumptions about the technology are being re-evaluated. As it stands, only a fraction of organizations receive the full benefits of RPA (Willcocks, 2019), and public sector organizations who are jumping on the RPA train now would therefore benefit greatly from studying the past failures of private sector organizations and adopting a more holistic and strategic approach to their RPA initiatives. In this chapter, I have detailed four areas that public sector organizations need to consider when dealing with RPA (see Table 1 for the summary).

RPA carries not only challenges but also plenty of opportunities. Hype for the basic RPA technology might be dying, but hype surrounding the generative potential of RPA is alive and well. There are intensifying discussions about combining RPA and elements of machine learning to achieve so-called intelligent automation (Coombs et al., 2020; Lacity & Willcocks, 2021), which would be capable of input analysis and limited decision-making, and researchers envisage solutions that combine chatbots and RPA for use in customer service and healthcare (Anagnoste et al., 2021). For the public sector, RPA could serve not only as a means to increase efficiency, integrate legacy systems, and eliminate routine work but also as a foundation for innovation to improve work environments and offer better services to citizens.

Table 1 Summary of recommendations

1	RPA is easier to start than to scale	RPA is an attractive technology due to its low cost and fast deployment, but this does not mean that RPA is easy to scale. The first RPA projects need to be carefully selected, and organizations must get buy-in from key stakeholders and assign a champion to push RPA initiatives forward.
2	RPA requires long-term thinking	Organizations need to resist the temptation of the quick wins promised by RPA. Instead, they should ensure their approach to RPA aligns with the business strategy of the organization.
3	RPA is not all about FTEs	Traditionally, RPA is often evaluated in terms of potential cost savings brought by automating FTEs. However, there are increasing calls to focus on the broader value delivered by RPA, including workers reoriented from routine tasks that are suitable for automation and toward more value-generating and rewarding work.
4	RPA comes with accountability concerns	Public sector organizations manage a lot of citizen data and often make decisions affecting society. Such organizations, therefore, need to take extra care when integrating automation technologies into their organizational processes.

References

Anagnoste, S., Biclesanu, I., D'Ascenzo, F., & Savastano, M. (2021). The role of chatbots in end-to-end intelligent automation and future employment dynamics. In A. M. Dima & F. D'Ascenzo (Eds.), *Business revolution in a digital era* (pp. 287–302). Springer.

Asatiani, A., Malo, P., Nagbøl, P. R., et al. (2020a). Challenges of explaining the behavior of black-box AI systems. *MIS Quarterly Executive, 19*, 259–278. https://doi.org/10.17705/2msqe.00037

Asatiani, A., Malo, P., Nagbøl, P. R., et al. (2021). Sociotechnical envelopment of artificial intelligence: An approach to organizational deployment of inscrutable artificial intelligence systems. *Journal of the Association for Information Systems, 22*, 325–352. https://doi.org/10.17705/1jais.00664

Asatiani, A., & Penttinen, E. (2016). Turning robotic process automation into commercial success—Case OpusCapita. *Journal of Information Technology Teaching Cases, 6*, 1–8. https://doi.org/10.1057/jittc.2016.5

Asatiani, A., Penttinen, E., Rinta-Kahila, T., & Salovaara, A. (2019). Implementation of automation as distributed cognition in knowledge work organizations: Six recommendations for managers. In *40th International Conference on Information Systems*. Munich.

Asatiani, A., Penttinen, E., Ruissalo, J., & Salovaara, A. (2020b). Knowledge workers' reactions to a planned introduction of robotic process automation—Empirical evidence from an accounting firm. In R. Hirschheim, A. Heinzl, & J. Dibbern (Eds.), *Information systems outsourcing. Progress in IS* (pp. 413–452). Springer International Publishing.

Berryhill, J., Heang, K. K., Clogher, R., & McBride, K. (2019). *Hello, World: Artificial intelligence and its use in the public sector*. OECD.

Bygstad, B. (2016). Generative innovation: A comparison of lightweight and heavyweight IT. *Journal of Information Technology, 32*, 180–193. https://doi.org/10.1057/jit.2016.15

Cavoukian, A. (2009). *Privacy by design: The 7 foundational principles*. Inf Priv Comm.

Coombs, C., Hislop, D., Taneva, S. K., & Barnard, S. (2020). The strategic impacts of intelligent automation for knowledge and service work: An interdisciplinary review. *The Journal of Strategic Information Systems, 29*, 101600.

Costello, K., & Rimol, M. (2020). Gartner says worldwide robotic process automation software revenue to reach nearly $2 billion in 2021. In *Gartner*. https://www.gartner.com/en/newsroom/press-releases/2020-09-21-gartner-says-worldwide-robotic-process-automation-software-revenue-to-reach-nearly-2-billion-in-2021

European Union. (2016). Regulation (EU) 2016/679 of the European Parliament and the Council. *Official Journal of the European Union L, 119*, 1–88. https://doi.org/10.5771/9783845266190-974

Fersht, P. (2012). Greetings from Robotistan, outsourcing's cheapest new destination. In *Horses Sources Blog*. https://www.horsesforsources.com/robotistan_011112

Kroll, C., Bujak, A., Darius, V., et al. (2016). *Robotic process automation—Robots conquer business processes in back offices*. Capgemini.

Lacity, M., & Willcocks, L. (2015). What knowledge workers stand to gain from automation. In *Harv. Bus. Rev.* https://hbr.org/2015/06/what-knowledge-workers-stand-to-gain-from-automation

Lacity, M., & Willcocks, L. (2016). Robotic process automation at Telefónica O2. *MIS Quarterly Executive, 15*, 21–35.

Lacity, M., & Willcocks, L. (2021). Becoming strategic with intelligent automation. *MIS Quarterly Executive, 20*, 1–14. https://doi.org/10.17705/2msqe.00047

Lehtinen, J., Wiik, H., & Jalonen, P. (2020). *Your guide to speed and scale up enterprise-wide RPA initiatives in the cloud*. Capgemini.

Lindström, U.-M., & Sakslin, M. (2019). *Verohallinon automatisoitu päätöksentekomenettely ei täytä perustuslain vaatimuksia (EOAK/3379/2018)*. Parliamentary Ombudsman of Finland.

Martin, K. (2019). Designing ethical algorithms. *MIS Quarterly Executive, 18*, 129–142. https://doi.org/10.17705/2msqe.00012

Peña, D. (2021). Solving the RPA talent shortage with nearshoring. In *Auxis*. https://www.auxis.com/blog/rpa-talent-shortage
Ranerup, A., & Henriksen, H. Z. (2020). Digital discretion: Unpacking human and technological agency in automated decision making in Sweden's social services. *Social Science Computer Review*, 1–17. https://doi.org/10.1177/0894439320980434
Rinta-Kahila, T., Someh, I., Gillespie, N., et al. (2021). Algorithmic decision-making and system destructiveness: A case of automatic debt recovery. *European Journal of Information Systems*. https://doi.org/10.1080/0960085X.2021.1960905
Robbins, S. (2019). AI and the path to envelopment: Knowledge as a first step towards the responsible regulation and use of AI-powered machines. *AI and Society*.
Rosenfeld, A., & Richardson, A. (2019). *Explainability in human–agent systems*. Springer US.
Sicular, S., Hare, J., & Brant, K. (2019). Hype cycle for artificial intelligence, 2019. In *Gartner*. https://www.gartner.com/en/documents/3953603/hype-cycle-for-artificial-intelligence-2019
Syed, R., Suriadi, S., Adams, M., et al. (2020). Robotic process automation: Contemporary themes and challenges. *Computers in Industry, 115*. https://doi.org/10.1016/j.compind.2019.103162
Willcocks, L. (2019). RPA—The state of play today. In *Robot. Cogn. Autom.* http://www.roboticandcognitiveautomation.co.uk/Blog13.html
Willcocks, L. (2021). Automation and going digital Part 2: Is RPA yesterday's technology? In *Robot. Cogn. Autom.* http://www.roboticandcognitiveautomation.co.uk/Blog39.html
Willcocks, L., Hindle, J., & Lacity, M. (2019). *Keys to RPA success: How blue prism clients are gaining superior long-term business value*. Knowledge Capital Partners.

Index

G. Juell-Skielse et al. (eds.), *Service Automation in the Public Sector*, Progress in IS,
https://doi.org/10.1007/978-3-030-92644-1

The manufacturer's authorised representative in the EU is Springer Nature Customer Service Centre GmbH, Europaplatz 3, 69115 Heidelberg, Germany. If you have any concerns regarding our products, please contact ProductSafety@springernature.com

Printed and bound by CPI Group (UK) Ltd, Croydon, CR0 4YY

30/09/2026

02225497-0006